KU-440-312

Johnny Kingdom lives on the edge of Exmoor with his wife, Julie. For the past several years he has made a living running wildlife safaris on the moor and making videos of its flora and fauna. Johnny became a local celebrity with a nature series on ITV West but has now gained the status of a national treasure with his BBC2 series and Christmas specials.

For more information on Johnny Kingdom or to buy his DVD of Bambi, see his website at www.johnnykingdom.co.uk

Also by Johnny Kingdom

Johnny Kingdom – A Wild Life on Exmoor

and published by Corgi Books

BAMBI AND ME

A DEER IN THE FAMILY

Johnny Kingdom

CORGI BOOKS

TRANSWORLD PUBLISHERS
61–63 Uxbridge Road, London W5 5SA
A Random House Group Company
www.rbooks.co.uk

BAMBI AND ME
A CORGI BOOK: 9780552157346

First published in Great Britain
in 2008 by Bantam Press
a division of Transworld Publishers
Corgi edition published 2009
Copyright © Johnny Kingdom 2008

Johnny Kingdom has asserted his right under the Copyright, Designs
and Patents Act 1988 to be identified as the author of this work.

A CIP catalogue record for this book
is available from the British Library.

Addresses for Random House Group Ltd companies outside the UK
can be found at: www.randomhouse.co.uk

The Random House Group Ltd Reg. No. 954009
The Random House Group Limited supports The Forest Stewardship
Council (FSC), the leading international forest certification organisation.
All our titles that are printed on Greenpeace approved FSC certified paper
carry the FSC logo. Our paper procurement policy can be found at
www.rbooks.co.uk/environment

Typeset in 11.5/16pt Fairfield Light by Falcon Oast Graphic Art Ltd.

Printed in the UK by CPI Cox & Wyman, Reading, RG1 8EX.

2 4 6 8 10 9 7 5 3 1

Photographs courtesy of the author, Ralph and Jackie Rashley,
Rupert Smith, Steve Govier and Steve Guscott.

Every effort has been made to obtain the necessary permissions
with reference to copyright material, both illustrative and quoted. We
apologize for any omissions in this respect and will be pleased to make
the appropriate acknowledgements in any future edition.

This book is dedicated to our two
lovely granddaughters Roxy and Louise.
You both mean so much to us, you
are very special.

Contents

Acknowledgements

For all they gave Bambi, I'd like to thank:

All my family, especially my granddaughters, Roxy and Louise.

Brian and Lynn Buckingham of Polworthy Farm.

The vet Martin Prior and all his staff.

Gill and the late Roy Johnson.

Tony Niven.

The late Adrian Adams and his family.

The late Joe Drewer and his family.

Lynda, Justin, Jodie and Millie.

The villagers and visitors who brought treats for Bambi.

For the sawdust – Mike (known to me as Brother) and the carpenter Andrew Haydon.

For the pictures – Ralph and Jackie Rashley, Rupert Smith, Steve Govier and Steve Guscott.

1 Life begins

If I go through my kitchen now, and out the back door, into the garden, it's a different place. The smell is what I notice first. There's no more the scent of a deer, no more sacks of sawdust for her bedding, and no more of her number 18 calf cake that she liked so much. There's no more of her strange barks, the bumps and thuds as she moved around her shed, the excited cries of the children who visited her. Instead, there's just the calling of the birds, and the breeze in the treetops, and the odd car going past at the top of the road. In terms of the look of the place, well, the fence is gone. We simply took it down. There's a decent lawn now, the grass a bright, pale green because it's so fresh and new, without a single muddy footprint on it, whereas there used to be a sort of racetrack burned into the ground by Bambi's hooves as she raced around playing her various games. There's a pond we've made in the bottom left corner. We've planted new shrubs. There are neat stepping stones making a path across the grass. Most of all, there's space to move around. It's like a proper, normal back garden.

Because there's no Bambi.

Twelve years she lived with us, one of the family, and now she's gone. My wife Julie and I have two sons, Stuart and Craig, and we have our grandchildren, Roxy and Louise, and we had Bambi; she was part of the same scene. It feels very empty without her, truly it does – I'm not used to it yet. I'm going to get a lot of pleasure out of writing this book because there are so many memories, and I want to keep them safe, not let them go.

As I sit here with a notebook and a pencil, and I begin to scribble as many of them down as I can, the blackbird, Fred, and his young, swoop down to say hullo and feed on the raisins I bring them. The cat jumps up on the table here and rolls on her back. It's a rare sunny day in July of 2007, but I'm remembering all the way back to 1994, the year Bambi came into our life.

Of course I never saw her born, and I don't know what exactly happened but I've got a pretty good idea how it must have gone, before she and I met. After all, I've watched the deer up on those hills, day in, day out, wind and rain or in bright sun, in the autumn when the stags are rutting, or when the snow's falling, ever since I was a small child. And I'm nigh-on seventy years old, now, as I write this. That's a long time. And I didn't just watch them from a distance, through binoculars and using long lenses. I've been up so close, in full camouflage gear, that I've literally had to hold my breath. I've had a scare or two, thinking I was going to end up gored by an antler. I've gone three nights in a row up in the branches of a tree, just to get rare video footage of deer wallowing in a mud bath below. I've loved these creatures all my life, and I know every inch of the territory they make their own – that is to say, the broad, open-backed hills and steep wooded valleys of Exmoor. And I've often imagined, in my mind's eye, how it must have been for Bambi, in the few days before she and I happened to be thrown together by accident.

Polworthy Farm is a 185-acre farm on the southern slopes of Exmoor, just north of the river Yeo. Like Bambi, most deer calves are born in the first two weeks of June, and Bambi's mother would probably have given birth in the game crop that's next door to the farm. It's a stand of maize that's grown specifically to rear pheasants in, but it provides cover for all sorts of wildlife. Most likely first thing in the morning, she'd have separated herself off from the herd, stood apart – not too far away, mind – but she'd have become preoccupied with the signals coming from her belly that she was about to go into labour. I've watched this happen myself, many times. A hind leaves the herd and wanders away to find a place thick with ferns or bracken, or where the grass is very long. If it's grass she chooses, then she may come up against the problem of the early silage crop. A tractor driver mowing the early growth of grass, if the weather is right, can easily run into a red deer calf hidden away. I like to see tractor drivers carrying gloves in their cabs because if they move a baby calf using gloves, more likely than not the mother will come back to it, and take it on again just the same. But if the scent of mankind is on the calf, the mother will judge it's too dangerous and abandon it. Brian Buckingham – who farms at Polworthy – believes it was the height of the maize that Bambi's mother was attracted to, it's what made her feel safe. The others in the herd would have wandered on and left her to it. She'd have stood, dead quiet and still, mute, and after a while she'd have lain down, and she'd have made sure she was almost invisible. Even with all the pain of giving birth, she'd be wanting to join the herd again, just as soon as she could. She'd want to get the other side of this, and be up and able to run. A deer separated, and off its feet, knows it's in grave danger from predators.

The labour, if it was a normal one, would have taken maybe an hour. Bambi's mother would have kept as quiet as possible. You wouldn't have known anything was happening unless you were right there. The calf would have slithered out, wrapped in its mucus membrane, looking as if it were dead, just for the first few seconds, because it would lie so still. Her mother, quick as anything, would be on her feet, the afterbirth trailing from her back end, and she'd have swung round, lowered her head, and her first concern would be to lick around the mouth, nose and eyes of her baby calf. The rough surface of her tongue would have cleared the mucus membrane from Bambi's airway, and also it would have stimulated a reaction from this brand-new, very small calf's body. And so Bambi's brain, receiving the message sent by that sandpaper-like tongue moving roughly against the skin, would have sent out an answer, saying, 'Yes, OK, now's the time, I will take my first breath. I will join the world. I shall wake up. I'll open my eyes.'

And Bambi would have breathed, and lifted her head – and this is the bit that's most amazing, to us – she'd have probably thrown out a foreleg, and rolled on to her front, and sat up, just a minute or so after birth. Her mother's tongue would have been working down her backbone, stimulating the whole packet of nerves that are carried in the spinal cord, helping to massage life and warmth into her.

The most important, life-saving attribute that a deer has is to run fast – to run now. Within two minutes Bambi would have been up on her forelegs. The hind legs would take a bit longer – there's more of them to control – but within another minute Bambi would have been up and standing. A bit wobbly, fair enough, but a miracle of nature. Compare that to us. We take more or less a whole year to stand up and walk on our own.

When a calf first stands it looks so vulnerable. The legs are already two-thirds the length they'll be at full grown, so it would be like she was up on stilts, legs stuck out any old how as she got to grips with her strange new environment. Across her back would have been scattered the familiar white spots, the rest of her fur a golden russet colour.

Bambi would have staggered a bit, and the scent of her mother's milk would have drawn her quickly into having her first suckle. The life-giving colostrum, the rich early milk that provides the first dose of antibodies and the important early nutrients, would have been her first drink, thicker and darker in colour, different from the normal milk that's let down later.

No time would be wasted, now. Bambi's mother would know they were in danger. Every nerve in her body would be stretched to breaking point. She must find food. She'd probably have eaten her own afterbirth to take back in some of the nutrients she'd lost. But Bambi wouldn't have been strong enough yet to come with her to forage for food. So Bambi would have to be hidden, tucked away in a safe spot, while her mother went off. The spots on her back are fantastic camouflage. You'd practically have to step on her before you'd know she was there. She'd keep dead still, tucked away in a patch of undergrowth. Her mother would be some way off, maybe as much as a field distant. She might have been on her own, or with two or three other mums who'd just given birth. At the first sign of any trouble, she'd have barked. A fox can take a newborn deer calf, no problem. After giving her warning, Bambi's mother would have run in a different direction, to take the predator further away from the hiding place, rather like some birds fly away from their nests to lure predators away. I've even seen a hind running from a predator, her calf alongside, and she's pushed it with her nose into the bushes, even as they were running, to hide it, while she carries on, to take the scent in another direction.

In any case, when Bambi's mother would have come back, it would be time for Bambi to take a big drink of milk, before being left again. That was how it would have been for the first three or four days. I have seen, close up, that magical first time when a calf comes out of her hiding place. Then, Bambi's mother would have scented the air to find the direction which the herd had taken, and she'd head off. Bambi would have been especially vulnerable to attack, bumping alongside her flanks.

Bambi and her mother: a new family. For at least a year, they will stay together. Often you will see three generations in a family group: mother, daughter, and granddaughter. But the priority during these early months is the protection of the young. Only the safety of the herd can help them. Bambi and her mother must find the herd, and her mother must keep on eating grass to keep up her milk. Other than that, it looked like the dangerous part was over, for them.

Little would either of them have known what life held in store.

2 Two accidents

The night Bambi was born, it was life as normal for me. I worked in those years both as a grave-digger and a wildlife photographer, and I took people out on safaris to see the wildlife on Exmoor. I lived in the same house as I've always lived, since 1965, a council house in the little village of Bishop's Nympton, only a few miles from where Bambi had just been born.

I remember it was Julie's birthday, so I could make a good guess and say I might have shared a dinner with her and some of our friends down at the Bish Mill pub that evening, with a few pints of cider put away as well. The likelihood was that the same night that Bambi was born, I'd have been flat on my back, snoring, dreaming of the goal I'd nearly scored for the Bishop's Nympton B team. Only a few miles away, but a world apart.

I'd have woken at six, like normal. I'd have had my usual breakfast: a slice of bread and syrup, same as I've always had, ever since I was a boy. What would have faced me that day, as work? It might have been that I was taking a group of people on safari. It was a year or so after I'd had my first programme on national television. A documentary team from Yorkshire TV, directed by James Cutler, had found out about me and had followed me round, making a film called *The Secret of Happiness*. After it was shown I'd gone through a manic six months of being famous in a funny kind of way, opening country shows and so on, appearing on other television programmes, talking about my life, and selling hundreds of copies of my wildlife films, all sent out from my front room. But this wave of popularity had passed over and had gone, just as quick as it had come. However, it had left something behind. It had given me a boost, and I was keener than ever on photographing wildlife. And it also meant that I often took people out on the moors to help them see what could be seen.

Or, if I wasn't taking a safari, it could have been that I had a grave to dig, maybe along with my sons. It's a business that's run in the family for some while: my father and his father before him all dug graves.

I've done hundreds of them, I've seen countless people into the ground; ever since I was a small boy and helped my father. It was hard work until the mini-diggers came along, but they put most of us out of a job. It was how I started my career filming wildlife: I'd take a camera with me to the graveyards of Exmoor, and while I was digging I'd see what was going on in the ivy-clad walls of the churchyards, and in the trees hanging overhead, and I'd take a break and set up my video camera, and get all sorts of pictures, and then I'd cut together a little programme of my own using two video recorders, and show it in the local village halls and so on.

Anyway, it would have been that sort of day. A normal family day. Work, the wildlife, the folk around here. A pint of cider.

Whereas Bambi was about to suffer a terrible accident.

Up on the hill above Polworthy Farm there's a hedge that divides the game crop from the pasture field. And on the Polworthy Farm side of that hedge, it was fenced off. This is because sheep can scramble through a hedge, especially if the badgers or the deer have given them a head start and made a little path through. This gap in a hedge is called a rack, where I come from. To keep his sheep in, the farmer has to run a fence right up against the hedge, but the deer can jump these fences easily. So the racks are still there. And it's important, for deer, that they keep moving around. If they don't, inbreeding can become a problem. Not only that, of course, they have to go where the food is. During the summer, that might be out on the open moorland. In the autumn, they'll seek the shelter and food available in farmers' pastures. They like corn; and a herd of deer can eat a lot of it. Oats are their favourite. It's the richest food and good for growing antlers. They do a lot of damage to farmers' crops in that way. So, all in all, they do keep moving – during the rut a stag will move around in a circle about a mile in diameter, and he'll park his hinds all in a heap and then swing back and pick them up, move them on again. To win another group of females, he might travel as far as twenty miles. All this is to say that these gaps in the hedges, the racks, are a common sight, and it's an everyday thing for the deer to be passing through them, whether they're fenced or not.

Sometime during that day, Bambi's mother, with Bambi right at her side, approached this rack. She'd have been very hungry, that's for sure. Anyone who's breastfed a baby, or had a partner who's breastfed a baby, knows it takes a lot of extra food and drink to keep going. Because the body becomes like a machine when it has to produce enough milk to feed a young one. And all Bambi's mother had to do was pass through this rack and she'd reach the rich green pasture on the other side – just what she needed. She needed that grass to feed Bambi, who was bumping along behind her. Besides, this was her routine, anyway. She'd have known this rack of old. She'd have gone through here loads of times. Also, the rest of the herd might have gone that way, so she might have been aiming to catch up with them. Safety in numbers.

She'd have gone through the rack, and come up against the sheep fence.

Deer have an incredible jump on them. They can leap a five-bar gate without touching. If you want to keep them in – or out – you have to commission a special deer-proof fence, eight foot high. This was no doubt a fence she'd jumped many times before. And it was only a little scrap of a fence. For her it was just a little hop, the same as stepping over a kerbstone for us. She wouldn't have given it a second thought. She'd have gone over in one graceful movement. She started eating that grass.

For Bambi, coming along right behind, the fence wasn't too much of a problem. She should have been able to jump over it like her mother. Which is what she tried to do. However, something went wrong. Maybe she didn't jump but tried to scramble through it instead. Anyway, it's difficult to see how, but her offside back leg slipped through the wire, and the wire got hold of her.

She struggled. The wire twisted. She pulled – and the wire really caught hold of her, now. It tightened around her leg. She pulled harder; the wire gripped even more tightly.

Bambi didn't make it. She was hung up in the fence.

Her mother would have turned back at her calf's first call of distress. She'd have trotted back to watch over her. It must have been a painful sight. The mother would have trotted back and forth, calling anxiously but unable to help. If only someone had been right there it would have been a simple thing to help Bambi escape, but there wasn't any good luck around for Bambi, just then.

Instead she grew more tired, and the frantic kicking would have slowed down. She became weaker.

Bambi's mother would have walked back and forth a while longer. She'd have sniffed and licked her calf and called out, 'Mrrrrr!' It's a dreadful thing to imagine their distress. To be separated from the herd is like a crime, for them. Right next to one another, yet helpless.

It's difficult to guess how long it would have taken, but there'd have come the moment when Bambi's mother gave up hope.

Let no one tell you animals don't feel grief. They do, just the same as us. I've seen it countless times. Among the birds who've lost their young. All types of creatures. When they lose their babies unnaturally it's heartbreaking to them. I hate to think of Bambi's mother having to walk away from that scene. She'd lost her calf, but if she waited around, she herself would be prey. I can picture how it would have happened. She'd have trotted off, then trotted back for one last try, then wandered half-heartedly away again, come back. Eventually she'd have run away, mourning her lost calf.

Bambi was left alone. The pain must have been terrible. Exhaustion set in. The figure of death, with his scythe, a figure whom I've seen many a time while digging graves, was standing right next to her at that moment.

It's the same figure whose breath I felt on my own neck, once, when I too had a terrible accident. Those who've read my life story will know what happened, but it's worth giving a short account of it here, because it changed who I was and how I lived my life. Before the accident I was known for shooting deer, not saving them. I grew up in the war when food was scarce. From my earliest years I was trapping moles to sell their skins. I was sneaking after the rabbit man, making sure he couldn't see me, so I found out where his traps were and then I could take one or two before he got back round to them, and in our kitchen they'd be laid out and skinned, and put in the stewing pot. It wasn't just meat Father and I went after. We spent hours collecting wortleberries. Father kept a vegetable garden and used every trick in the book to grow his vegetables larger, better, and more plentiful. We had outside toilets in those days and it was my job as the only boy among six children to take out the foul-smelling bucket and dig it into the vegetable patch. In the next row of outhouses along, the venison was butchered. Father was infamous for his poaching, and I became a poacher just like him. My earliest memories are of the knives flashing in that outhouse, the smell of blood and fur, the meat being cut up. And the salmon being taken out of the rough old West Of England sack and gutted, divided up. We kept our friends and family in meat from killing deer and taking salmon from the rivers. This was our way of life; this was how us six children were fed when there was only a very low wage coming in from my father's work in the quarry. We were up to our elbows in the blood of animals, but no less did we love those same animals, and take pleasure in their beauty, and we were rightly proud of our knowledge of how they lived, and where they lived, and how to make the most of them, and how to steal them out from under the noses of landowners.

As I grew up, I graduated on to the bigger poaching expeditions. I became the one the police were after. They would love to have caught me. They tried, often enough. Once I was so angry with a policeman when he stopped me to search inside my van, that I would dearly love to have pushed him into the back so he could keep company with the carcass of a deer I had in there. I'd never do such a thing. So you wouldn't believe it, would you, when they threw me in the police cell, and they made me carry the deer carcass in there with me, and we had to spend the whole night in there, the two of us together, me and the deer my mate had killed. It was a time of midnight sorties with poaching lamps, of guns poking out of the side windows of moving cars like we were gangsters, a time of drinking and fighting and crashing cars. Those were wild days.

And then I had the accident.

I was working as a lumberjack, felling trees, but it wasn't a chainsaw that got me. I don't actually remember what happened: the first thing I knew was that I opened my eyes and something was licking my face – it was my own dog, Sandy, an Alsatian-lurcher cross. He was whining. I was lying in the mud. I'd been out for I didn't know how long. There was no one else with me. My eyes were open; I could move a bit, which meant that my arms and legs weren't broken. I didn't even know where I was. There was no tarmac or road around, so it couldn't have been another car crash. I was quite used to having those. When I tried to sit up I was sick and dizzy and veered close to losing consciousness again. The injury was to my head: I could feel one side of my face was terribly swollen and blood was pouring from my nose. My cheekbone was smashed in. My jaw was all wrong; it hung down like someone had unhooked it. I was in serious trouble.

I became more aware of my surroundings. The Mazda pick-up was right there, loaded up with logs. The tractor was not far off, and the four-acre cover of oaks, at Cowley Bridge near Exeter, that we were meant to be clearing.

So I was at work. But where was the man I

worked with, Robin Dyer? I remembered: today he'd had to go on a different job and I'd come up here on my own. I looked around. There was no house, no one else working here, not for miles. I was on my own. I had to get into the pick-up and drive. I made it to the door of the cab and hauled myself in. I was in terrible pain and my co-ordination was shot to hell and I was near unconscious from one moment to the next.

This confusion and pain must have been how Bambi felt, after she'd been hung up in that fence.

The journey in my pick-up was a blur; somehow I made it. A couple of times I had to veer to the side of the road because I lost consciousness again. Then I was back at Bishop's Nympton, standing in front of my own front door. The door opened and Julie was standing there. I saw the shock hit her; the blood drained from her face; she went white as a sheet. She caught me as I stumbled forward.

Julie had our two little boys to think of, so it was Gordon Parker who drove me to the doctor. At Barnstaple hospital they really had to set to work. Dr Buchanan said I was like a vehicle in a crash-repair shop – and it took him a long time to fix me.

Meanwhile I could begin to piece together what had happened. Tree-felling is dangerous work and you should always work in pairs. I'd been on my own. I'd cut down an oak, trimmed the branches and cut off the top. OK so far. Then I'd jumped into the tractor and started reversing, ready to hook it up to the stick and haul it away. As I reversed, I looked over my right shoulder to see where I was going.

At the rear of the tractor was a big, heavy anchor, made of iron, which we put down when we needed to pull the trees over. When the tractor was moving, the anchor was held up off the ground by a hydraulic arm, hooked up to a heavy-duty chain, each link about as thick as a finger. One of the links in that chain had broken. It was corroded or there was a flaw in the weld. The heavy iron anchor had dropped to the ground and stalled the tractor. At the same time the hydraulic arm – a rod of heavy-gauge steel around four foot long and two inches thick – flew the other way, upwards, into the cab. It moved quicker than the eye could see and hit me in the cheekbone, just an inch below the eye.

I went through quite a few operations to put me right, and all the teeth in that side of my jaw went rotten and had to be taken out, nerves drilled out and all.

My accident knocked me for six. My nerve went. I lost my confidence totally. I couldn't look at a chainsaw or a lumber yard. I couldn't look my wife or children in the eye, I felt so bad. I was prescribed Valium. I went from a man at the top of his game, having just bought his own home, to being off work on twenty-one weeks' statutory sick pay of £21.90 per week. There was no picking myself up and dusting myself down, like after other accidents I'd had. However much I wanted to, I found it wasn't possible this time. My nerve had snapped as surely as the chain itself; it wouldn't do the work, wouldn't lift me. I was wretched; I truly was down. I couldn't even go into a pub. I sat at home, my head in my hands.

In the end it was a certain Reverend Pennington who helped me, and the very special way in which he brought me back from that experience was perhaps the thing that made me a different person, and made me able to care for Bambi after her accident. I'd had a bad accident, just like Bambi. And I knew what it took to get yourself mended after an experience like that. It takes patience, and tenderness, and loving care. With the Revd Pennington it was the same – it was in his nature to offer people a mixture of kindness and understanding, and patience. Also, he was interested in the whole picture. He helped me to give up smoking. He diagnosed the fact that I hadn't slept properly, deeply, since I was a small child, due to my sleep-walking. It was like having someone walk all the way round you and take care of every bit.

And – although I didn't know it yet – I would have to do the same with Bambi. It was her turn; she was on her own with no one to help, and that same figure of death loomed over her just as it had done over me. I would have to save Bambi, just like I'd been saved, myself.

3 The rescue

The first bit of luck that Bambi had – and thank God for it – was the highly tuned sense of smell belonging to a dog called Merc. He was a working collie owned by Brian Buckingham, who together with his wife Lynne runs Polworthy Farm. It's a mixed livestock operation: cattle and sheep. Looking down from the moor above, its beauty is beyond words.

At that time of year, early June, it's Brian's custom to drive the quad bike round to check his sheep at around 8.30 a.m. That morning, same as any other, he went on up to make sure the water troughs were clean and full, that no sheep was lame or ill, and so on. Alongside him, or sometimes in the dog box on the back of the quad, was his sheepdog, Merc. I presume he was named after the brand of car.

Maybe he's a fast dog, and good at going round corners.

Anyway, there wasn't another soul up there, and it can be spectacular on that high ground, near enough seven hundred feet up, with Exmoor just a field or two away to the north, and the tors of Dartmoor visible through the early morning mist to the south.

It was Merc who scented her. Brian tells me that Merc suddenly put up his nose and started to make a beeline in the direction of that rack. Brian didn't know what was up but he steered the quad to follow Merc and find out. He suspected it was going to be a dead sheep, or something like that. As he came near the hedge he saw the smudge of colour, rust flecked with white dots, where there should have been only the lines of the fence. It was like a piece of washing blown into it. And it was hanging upside down, struggling like mad. He came to a halt and switched off the quad. That's when he heard Bambi call out. It's a strange sound, a deer calf's cry. It sounds a bit like the 'meow' of a cat. He hurried over, the picture becoming more clear with each step. Bambi, only a day or two old, hanging there. The terror had driven all hope out of her. Brian tried to unwind the fence from around her leg but there was no way, it was too tight.

Among the kindest things he might have done, then and there, was knock her on the head and help her escape the suffering, since it was almost certain she was going to die anyway, given she'd been abandoned by her mother. He could have left the carcass to be pulled apart by foxes, crows, buzzards, in the natural way of things, and that would have been Bambi's fate. Yet something in him meant he couldn't do it, couldn't turn his hand to it. Red deer calves are incredibly beautiful things, so dainty and pretty. A sympathy for Bambi, and a wish that something might be done, that she could be helped back on her feet and released back into the wild, meant that instead of putting her out of her misery he jumped on the quad and burned on back to the farm to fetch wire cutters. There and back was only a few minutes. A couple of seconds later she was cut free.

Brian knew Bambi's mother had abandoned her, so he drove the quad slowly, one-handed, back to the farm, cradling Bambi in the other arm, with Merc running alongside. He brought her into the yard and put her down in an unused stable. It was dry and warm in there, with a bed of straw. If some milk could be got down her, if she could be kept alive for between six and nine months, and if she could be weaned successfully, there was a good chance of returning her to her family. He didn't think the injury looked too bad. After all, she was walking around OK, she was putting weight on the leg even if she did have a heavy limp. He went in and fetched his wife, Lynne, and the children, and they all came to see what a beautiful creature was suddenly in their yard.

What should they do with her? She'd survived this long, which was something of a miracle. Should they call the RSPCA? Or keep her and try and feed her themselves until she was grown enough to be taken back out and let go? As they worked out the possible alternatives they remembered that I had a great love of animals, and had looked after all sorts of orphans and injured birds and so on, and I knew the lives of deer up on the moorland better than most, and so they thought it was worth talking to me. Brian went in and phoned me. I remember the conversation very well.

'Johnny, it's Brian Buckingham here.'

'Brian, hullo.' I wondered what was up.

'Guess what's happened, I found this deer calf caught in the fence, you know, up at the top?'

'I know, oh yes.'

'She can't be more than a few days old. I brought her down to the yard and made up a little stable for her. She's up and walking around, she doesn't look too bad.'

'Where's the damage, then?'

'It's her back leg. The wire was tight around it, but she's putting weight on it OK. And I was thinking, who's going to care enough about her to see her right, put her back to where she ought to be, and I thought of you.'

My heart leaped to my mouth at these words. I did feel a bit choked up. I used to be the kind of person someone called if they wanted a deer killed for the pot; now I was the first man Brian thought of who might save this little creature, nurse it back to health. I couldn't hurt another creature, not after my own accident. Brian's words were evidence of how much I'd changed. Given everything that had happened, the injuries I'd suffered and the sort of mental breakdown, really, that had gone with it, here was proof that I was out the other side of it, a different person.

'We were wondering,' went on Brian, 'if you'd care to come up and have a look at her, and then we can see what we want to do?'

'Course I will, right now.'

'See you later, then.'

I went up to Polworthy Farm straightaway. I had

a Daihatsu jeep at that time, which I'd got as part of a sponsorship deal connected with the television programme. So I was in that Daihatsu, not having a clue what I was letting myself in for. After about four or five miles of country lanes I turned down the rough track to the farm and pulled up in the yard. There are a lot of buildings up at Polworthy and at that time of year, early June, there's many different things going on with a farm that size, so there was the noise of lambs in the lambing sheds, calves and so on. Everyone was busy; there was no sign of them. I looked around, calling out for Brian and checking this barn and that, and, finally I came across her – there she was, in a stable of her own.

My first thought was: God, she's so tiny, standing there. Just a scrap of a thing. Only a few days old. A baby. What a shame . . .

As the little calf limped around, my second thought was the same one I'd had a thousand times when I'd been up on the moor, photographing her breed at this time of year: She's beautiful, what a creature, God's gift. The limp was heartbreaking. You know how it is yourself, when you get that bite of pain and you have to hurry the weight off it, quick as you can, to reach the safety of the other leg.

Brian joined me now. We went into the stable and

together we held her while we had a look at her injury.

'What d'you reckon, do you think she's got a chance?' I asked him.

'It's not broken. That's the main thing. So I reckon it'll mend, no trouble.'

I already knew I was going to have a go at getting her through this. I knew I was facing up to a lot of work and expense, but the thought of her running off, joining a herd, mended, meant it was an easy decision. There was a chance we could do it and it was worth taking that chance. We'd see her right, until she was strong enough to go back to her own kind. I stroked my hand along her back. The fur was so fine and glossy. Her eye was bright; she was in shock. The pain must have been bad. 'I'll take her with me,' I said. 'And it will be one step at a time and see what happens.'

I backed the Daihatsu jeep up to the stable and lifted her into the front seat. She was too small to go in the back; she'd just slide around. She didn't struggle at all, as I remember. As I drove slowly up the bumpy track I glanced over at her, curled up on the passenger seat next to me. Tiny little thing, she was! I could see her nose twitching. They all do that. Deer have an incredible sense of smell and you can watch their noses working, like a dog's. She was taking in my human smell. Every instinct would be telling her I was the enemy. The other thing I noticed was her eyes were glassy. Fright, that was. I'd have to watch out. A deer can die of fright quite easily and she was only a baby.

How odd, how strange, to be driving along the Devon lanes as calmly and contentedly as if we were old friends, her looking out the windscreen – but my companion was a deer calf.

And she got her name, on that first journey. I didn't exactly have to think about it. Bambi, of course. It just sort of stuck in my head and didn't go away. She was called Bambi.

I drove slowly and carefully the four or five miles back home and parked the Daihatsu out the front of the house. I went in and called Julie to come and have a look.

We've been married for nearly forty-five years, Julie and I. I first laid eyes on her when she was thirteen years old, although I don't remember it. It wasn't until she was fourteen that I first noticed her. Her family was against me because I was into fighting and drank too much cider, but we managed to walk and talk. I went away and did my National Service in Hong Kong, but when I came back the only gift I brought with me was for her, and we took up again where we left off. I managed to win over her father by bringing him gifts of fresh salmon, but I could never tell him about shooting deer. Julie said yes, she would marry me, and that was in 1963. We've been together ever since, although anyone who's read my life story will know there's been a time or two when she must have wished she hadn't said yes, that day. We have had two sons together and, like I said in the dedication of my last book, she needs a handful of gold medals; she is everything to me and has been right from the start.

And Julie loves animals as much as I do.

'Oh, the dear little thing,' she said when she first saw Bambi. But like me she was very worried. If we thought the worst, then this creature wouldn't survive. There was a strong chance that all we were doing was letting ourselves in for serious heartache, watching her die, and a lot of work for nothing. 'You know what it's going to mean, don't you?' I said. Talking half to myself, I was. 'Feeding every few hours, up all night. And if she lives, we're going to have to look after her for as long as six months before we can turn her out again.'

Julie knew as well as I did that it was going to be a lot of hard work. For six whole months. Little did we know! Bambi wasn't going to be a temporary visitor, she was going to become a member of our family.

But we were on this track, now. We'd taken this on. We were going to give it our best shot.

We carried her in through the front door, into our dining room. It was in here, not even a year ago, that the entire family had sat filling envelopes with videos – hundreds of them – in the weeks after the TV programme had appeared. And on those videos had been all the best shots I'd got of deer in the wild, close up, gathered over many years. Shots of herds twenty or thirty strong, cropping grass on the moorland hills. A famous shot of a whole herd of deer wallowing in a mud bath while I hid in the branches of the tree right above them. Deer crossing rivers. Stags fighting, lowering their heads and charging one another, antlers clashing like crossed swords, during the rut. A calf hidden so cleverly in a patch of nettles you could hardly make her out, even though the camera was only a few feet away. A lone stag, agitated, on the crest of the hill, having scented me and not quite sure what to do about it. A stag nosing past, in deep woodland, growing his antlers so fast you could almost watch them get bigger in front of your very eyes.

And now, here was a real one, her life hanging by a thread, being carried in. What must Bambi have thought of it all? The table, the TV, the china animals in the glass-fronted cabinet? Lord only knows how strange it was for her. We took her through the kitchen, out the back door and into the back garden, which was Julie's domain.

She'd done a lovely garden, had Julie. Everyone remarked on it. Full of flowers and shrubs, a sunny place. When we'd first moved in here, in the mid sixties, it was a bed of stinging nettles. Her father and I had cleared it, and it took weeks. Next it was a veg garden. When the children arrived, there was a lawn put down for them to play on. Down the bottom was always a bit of a farm: chickens, ducks, pheasants, ferrets. I planted some trees. After the animals had gone, Julie had made the garden. Most of the shrubs had been given to her by various friends and neighbours. Roses, irises, lilies. She could walk around that garden and she could point at every plant and say who'd given it to her, so it was like the person was always remembered. It was hard work because the ground wasn't too good. She'd made us a lovely garden, and she was rightly proud of it. Often our grandchild Roxy ran around in there. There was plenty of room.

Quite close to the house is the garden shed. It's not one of those pre-fabricated ones you can buy at a garden centre. It's put together out of rough straps of timber, kind of like a scout hut or a log cabin. This was where we intended to put Bambi; we'd make a home for her in there. We pulled open the door and of course it was full to the brim with rakes and forks and shovels and pots and pans and watering cans and buckets and bits of old gardening equipment. My son Craig's old spacehopper was in there. His daughter Roxy had a toy wheelbarrow she played with when she came round and that was in there as well. I waded in and piled some of the stuff up to one side so we could make a little space for Bambi. I put down some kind of blanket for her to lie on. It's in the photograph – you can see how cramped it was. We just had to make a safe, warm place for her, straightaway. I carried her into the shed and found another blanket – I think we pulled it out from between the spokes of an old bicycle, but it did the job, a nice blue checked colour – and I laid that over the top of her, to keep her warm.

The first thing on our minds was to get some warm milk in her, and to call the vet. I telephoned Martin Prior straightaway and told him what had happened. His response was immediate: he'd come over and take a look.

Just as urgent was to find a source of milk. She'd been some time without and a calf this young would normally be fed every few hours. Ideally it would have been milk from a deer, same as her mum's, but of course there was no way of collecting any of that. It simply wasn't available. Some people do farm deer, of course, but they never go near them, they're bred for meat and there's no way of milking them. Cows' milk was plentiful, any farm around here would be able to give us cows' milk, but it wasn't any good for our purposes. The lactose content is too high for a deer calf; it would be very difficult for her to digest. It would be like poisoning her, to give her cows' milk.

Thank God for Jill Woollacott, as she was called then. She kept a herd of goats. And goats' milk was what Bambi needed. It was the nearest thing.

When it came to writing this book I called up Jill to see what she remembered, and it turned out she remembered everything, the whole scene.

'Oh yes,' she said. 'You called me up one day, Johnny, and you told me you'd picked up a baby deer calf, and that goats' milk was the nearest thing to its mother's milk, being so very rich but without the lactose. You asked if you could come up. And then you arrived, and you had Bambi with you, in the truck. She was ever such a beautiful little thing. She had all her legs then . . .'

I carried Bambi into her barn, and as it happened Jill had a particularly quiet goat, who was called Diesel, which wasn't the most feminine name, but apparently she was a prolific milker and, more to the point, Jill said, she was a very peaceful goat. She was a Saanen, one of the bigger breeds. At that time Jill had sixteen of them, tethered on bits and pieces of rough ground about the place – it's the rough ground they like. She'd unhook their tethers and lead them into the barn two at a time, and milk them. She'd put a piece of hardboard up on two blocks, and she'd have the bucket up there on what amounted to like a bridge or a platform affair, and the goat was then high enough for her to put down her milking stool and milk the goats without having to stoop, which saved her back. Meanwhile the goat would eat its breakfast from the bucket. And as it happened, Diesel was up there when we arrived. So I carried Bambi in, and we were both quiet as mice and moved slowly, and we brought Bambi right up alongside Diesel, and sure enough, it didn't take long before Bambi was suckling off this goat, and the goat wasn't the slightest bit worried. Jill and I were both amazed at the sight of that. It was really touching, this little deer calf with its terrible limp, and so hungry and far from where it should be, sucking from one of Jill's tame goats. It brought a tear to your eye to see that. It felt like huge progress had been made, in one stroke. Good old Diesel, I thought. And I was shaking my head a bit at that point because I'd wished I'd brought my camera along to film it happening, it was such an odd sight. And Diesel didn't mind a bit, she just accepted this as if it were the most natural thing in the world. I suppose when she was up on that platform she always had her milk taken off her, so it was business as usual, except it was a deer calf taking it instead of Jill.

I did eventually take my camera down to Jill's, another time, and I put it in the film I made about Bambi. It's amazing: that old barn of Jill's looks like an oil painting done centuries ago, or like a scene from one of those BBC classic films, with a gentle light washing over the old cob walls, and in the stall a goat stands, with Jill in her striped apron there, pulling the stool up close and milking it, for all the world as if there were no machines, no electricity, no motor cars nor running water and we've all been thrown back into history. Except a moment later I step into the shot, wearing a bright coloured shirt with a flared collar, so we're in the twentieth century again, suddenly. There's a gaggle of us watching her.

We talked about Bambi staying with Jill, and Bambi having a goat for a mother – maybe Diesel could be her mother? That was a real possibility. It was obvious to us it could have worked. It would have been a different life for Bambi. But we couldn't think too far ahead, because first of all we had to think about the vet and getting that leg seen to.

Because her leg was worse; she needed the vet urgently.

Me with Bambi in the back garden just after she was found. As you can see, she still has her back leg. She was such a beautiful, dainty little thing.

Above: *When we first brought Bambi home we settled her in the garden shed – you can see all the rubbish we kept in there at the time. Roxy was ever so good at feeding her, even though she was only tiny herself.*

Below: *Lying down in our garden with her injured back leg, Bambi was just like you might find a red deer calf in the wild.*

Above: *We tried to keep Bambi warm and safe in our garden shed but you can she how shaken she was by her injury and finding herself in such a strange place.*

Below: *Me with our neighbour Mark Woolcott feeding Bambi. Right from the very beginning our friends came to help.*

Above: *Little Roxy helps me to feed Bambi. I had to hold the bottle too as Bambi was so keen to suck on it she almost pulled it out of Roxy's hands.*

Below: *My wife, Julie, taking Louise to see Bambi when they were both very young, but very interested in each other.*

Above: *I took this picture when I'd been trying to film Bambi. She was fascinated by the furry microphone cover and kept trying to eat it.*

Below: *Me introducing my cousin Denise and her son to Bambi.*

Above: *Both my granddaughters, Roxy and Louise, loved to visit Bambi.*

Left: *Me and Bambi both looking pretty shaggy. Bambi was about to lose her winter fur and I was about to shave off my beard (my wife wasn't very keen on it).*

Below: *Bambi posing for the camera in the garden.*

Above: *With Roxy and Julie, Bambi was very much part of the family.*

Below: *Bambi with our neighbour Lynda from over the road. Julie looked after Lynda when she was little (Lynda's children, Jodie and Millie, even call Julie 'Nana') and in her turn Lynda looked after Bambi when we were away.*

Above: *In the summer I used to give Bambi showers. Sometimes I would hang the hosepipe over the branch of a tree and watch her run in and out of the water, having fun and keeping cool.*

Above: *Bambi climbing on to a bale of hay. She loved to box with it and play games.*

Right: *An alternative to a shower was the water that Roxy and Louise fired out of their water pistols.*

Below: *I practise my filming technique on Bambi from behind a tree. I think she's spotted me and is coming for my microphone.*

4 A dangerous operation

The veterinary practice in South Molton has a generous attitude, as far as wild animals are concerned. If they hear about one that's in trouble and needs attention, generally speaking they will treat it for free. It would be quite easy for them to say no, or to charge the person who makes the call, but they don't. Martin Prior put it like this, when we were talking about it the other day. He's looking back now on quite a number of years of being a vet, and he said, 'It's like a privilege, though, isn't it? We are lucky enough to be able to help and that is our good fortune.'

Martin Prior came and looked at Bambi. He was very gentle. He talked to her and checked her mouth and so on, and then he took a close look at the injury. The fur below the site of the wound was an odd grey colour, and duller. It looked different. He told us that the blood supply to the lower leg had been compromised. The wire being tight around it for so long had clamped the arteries and veins, and sealed them shut. As he explained, as soon as you cut off the blood supply to any area of the body, it begins to die. And when flesh dies, bacteria immediately begin to breed, and the flesh begins to rot. Just up from the wound, for a few inches, it was swollen and hot. Bambi's body was fighting back against the toxins created by the poisoned flesh below, trying to prevent them from spreading around the body, carried in the bloodstream. Her own circulation was going to be her worst enemy.

She was in trouble. I could tell from his tone of voice.

He gave her an immediate dose of antibiotics to help her in that battle with the poison, as well as painkillers to ease her suffering. However, there was no getting away from it: the blood supply to the bottom half of her leg couldn't be repaired. It was too late. The damage had been done while she'd been up there, trapped by the wire. Amputation was the only answer. And if we didn't act quickly, the poisoning would intensify and overwhelm Bambi's antibodies, and the toxins would quickly travel around the body and destroy all her major organs. She would die of septicaemia.

The date was set for the operation: tomorrow morning. She was to be brought in tonight for prep, and to be kept off food. It was important her stomach was empty, but she could be kept going on a drip. We drove her to the surgery and carried her in. They take such good care of animals in those places, but it breaks your heart to see all the different creatures in their cages, and the looks on their faces saying to you, 'Get me out of here, this is a terrible place, this is torture.' It's so sad that you can't explain to them the kindness that is being done in there.

Martin Prior could see that I'd already bonded with Bambi. Yes, I was worried, and I was right to be. Martin Prior had told me honestly that he reckoned she only had a twenty-five per cent chance of survival. She'd be in a weakened state from lack of food and water. The drip he put in, to carry water and energy directly into her veins, bypassing all the chewing and digesting necessary with food, that would help, but she was ever so young and tiny and frightened. I was too close to her to be able to watch. So, like the relative who has to wait outside, I went home and waited by the phone while Bambi herself waited ... She would undergo the operation first thing the following morning.

I spent the night pacing round. I couldn't get that poor deer calf out of my head. I brushed my teeth – so many of them false on the right hand side, having lost them all in the accident. It was an odd feeling. I was so used to putting on camouflage and visiting the world of the deer, almost living with them, but now it was Bambi's turn to visit our world. It was unfair that I had got so much pleasure from living on her side of the tracks, whereas she was having a horrible time over on our side. I put on my pyjamas and went through all the usual nonsense to do with going to bed, but I hardly slept a wink. I kept on thinking of that poor creature. How lonely she must be feeling. How strange for her, just the fact there was a cage around her. That she was visited by human beings, whom every instinct would be telling her to flee from, and yet here they were, taking her in motor cars, carrying her in and out of houses, taking her to South Molton, sticking needles in her, putting her in prison, giving her strange-tasting food to eat.

I lay down and turned out the lights and told myself to go to sleep, but then I thought of the pain she was in from that leg, and I remembered the terrible pain and confusion from my own accident, and I sat bolt upright. I climbed out of bed and looked out of the window. Over in that direction, a few miles away, the deer calf was maybe awake too. I could imagine the unblinking darkness of her eye. I wanted to send her a message: Hold on, we're coming for you. Just hold on. We'll see you all right. We'll look after you. We'll mend you if we can.

Deer are an amazing breed of animal. They have physical strength and grace, combined with a strong herd and a strong family instinct. Yet they are delicate and timid, and live in such fear of predators – especially man – that they must keep hidden as much as possible. They are a rich part of this country's heritage, with a history that's always been interwoven with the fate of the royal family, with whom it is so closely associated. In the past, Kings of England have seized vast areas of the country just so as to have the hunting of deer to themselves, and any commoner caught with a deer carcass would have an arm cut off, or some other such cruel punishment. Yet it's been the hunting fraternity, you could say, who have saved the deer. Certainly on Exmoor deer would have been made extinct hundreds of years ago, at the time that sheep farming was so lucrative and any animals which competed for the same keep were unceremoniously killed. It was the huntsmen who organized the deers' rescue and kept the herds safe.

The next morning, while I was spreading syrup on to my bread and butter, Bambi went under the knife. Martin Prior later gave me a blow-by-blow account of the operation.

The first thing he noticed was that the dead flesh below the site of the wound smelled even worse. 'It's like black udder,' he said, which is a disease cows get. The scent is unforgettable, and one that all of us know by instinct, and recoil from – because it means death.

Bambi was lifted on to the operating table – a steel affair that moves up and down – and she was anaesthetized. This was the most dangerous part of the operation. It's like flying an aeroplane: the difficult moments, when there's most stress on all the machinery, on all the systems, if you like, are the take-off and landing. The veterinary nurse was in charge of the procedure. A gas mask was put over Bambi's nose and mouth and she inhaled a gas called halothane. When she was unconscious a tube was put down her airway. This is particularly tricky for a ruminant, which is the class of animal a deer comes under, the same as cows. Ruminants gather grass while the going is good – say while the weather's OK and they can see there are no predators – and they put this freshly cut grass in a first stomach – you can think of it like a store or a larder. It's the same as a crop, on a pigeon. Later on – often in the evening – they bring this grass back up again into their mouths to chew it, to ruminate. It's a tough job, at the best of times, to get any nutrients out of grass and the ruminants need all this machinery to do the job. And then it goes back down, into their proper stomach. All this is to say it's a natural reflex of the ruminant to bring up stuff from their stomach, and there's a real danger they'll choke themselves if you go poking tubes down their windpipes when they're unconscious.

That part was safely done; apparently there were a number of other vets from the same practice who'd come in to watch this operation and everyone breathed a sigh of relief.

Now that the tube was in position a little collar was inflated which swelled the tube inside the airway and sealed it off, so now Bambi's breathing was controlled by the anaesthetist – the correct mixture of the halothane and oxygen came down that pipe, to keep Bambi alive and breathing but at the same time to make sure she stayed asleep. The drip was in position also.

The skin on her rump was shaved and doused in iodine.

Martin Prior explained to me that a human being with a lower limb injury can have the amputation only part of the way up the leg, below the knee, or just above the knee. But this is because a human can be trained to use a false limb, a prosthesis. Experience has shown that, with an animal, it's better if the whole leg comes off. If they take off half the leg, the animal will still try and use it, they'll try and put it down, and of course it's impossible. They adapt better if the whole limb is gone. They realign themselves; their balance is better. The bone was going to be cut across the mid shaft of the femur, right at the top.

The first cut, though – through the flesh – was lower down, just above her hock, her elbow. Martin used a scalpel to make a complete circle around the leg. More incisions were then made upwards, so the flesh could be peeled back. This spare flesh would be used to create a pad, a cushion, between the bone and the skin. This is because bone really hurts – like when you knock your shins, the pain's awful because the bone is near the surface. So they were going to wrap the spare flesh around the end of the bone to protect it. At the same time, as they pulled the flesh away, they teased out the major arteries and veins, and tracked them upwards. Just above the place where they were going to saw the bone they tied these arteries and veins off, using catgut. It's a miracle how you can tie off such major blood vessels and the blood just finds another way to go round the circuit.

Bambi was breathing easily through her tube. It was going well. The other vets watched carefully. Martin's next job was to saw through the bone. It's a practical kind of job, being a surgeon, like being a car mechanic. All sorts of systems to take account of: fuel, oil, electronics, and so on. Valves. Intake and exhaust. All sorts of tools for different jobs. You have to handle them pretty well. The importance of measurements. The fact that you have to get it right, or it's a disaster. The difference is – with surgery it's always a life-or-death situation.

Martin marked the point, set the teeth of the blade against the bone, and sawed off Bambi's rotten leg. It only took a few seconds but it saved her life. It was dropped into a waste bag and taken away and incinerated. He folded the spare flesh over the stump and stitched it in place to make the cushion. Next he closed the wound, sewing with a curved needle. The job was done.

Then came the most dangerous bit – taking her off the halothane and removing the breathing tube.

She was kept sedated, of course, but using the drip, now. This was because it was important she didn't move or struggle. The collar on the breathing tube was deflated and the tube itself pulled clear of her airway. It was a tense moment – but she was breathing normally on her own. Everyone gave a sigh of relief.

The operation had been a success. The whole thing had taken about an hour. Those who had been watching began to disperse. Martin called me to tell me it had gone OK. We were over the first hurdle.

The next challenge, for this little baby deer, was to recover. She was lifted off the steel operating table and placed inside a big walk-in dog cage, and laid down on a special soft, white, synthetic vet bed, which has the quality of allowing urine to drain away while the surface of the bed stays dry. And she was covered in bubble-wrap, believe it or not. Apparently it's the best insulation. On top of the bubble wrap came a blanket. She was only tiny, and the shock she had suffered could easily cause a dangerous drop in body temperature so she had to be kept warm. Through the drip she was kept lightly sedated and painkillers were also administered. We'd done all we could and, beyond this point, Nature would have to take its course. Martin told me over the phone that if

Bambi survived the night, it was probably going to be all right. All of us held our breath.

Once again I couldn't sleep very well. I kept on imagining the first person into the vets' premises the next day – who would it be, one of the vets, or a cleaner, or the receptionist? Whoever it was would get a first look at Bambi, and be able to say whether she'd survived the night . . .

You can imagine how much we waited on that phone call the next morning. It was a beautiful day, I remember. Julie and I wandered about, restlessly. We were out in the garden, pacing around. We passed each other in the kitchen, back and forth. Anything we talked about, and anything we did, was just to fill in the time.

The phone rang. I made a dive to answer it.

'She's made it through, she's fine,' said Martin Prior.

'Thank God!'

5 Bambi comes home

Once I'd shouted to Julie that everything was all right, Martin Prior went on to describe how he'd reduced the sedative, and Bambi had woken up. The first thing he'd done was to make sure the sucking reflex was active. It's one of the ways that vets can judge the wellbeing of very young animals. They can't talk, so you can't exactly ask them how they feel. But the sucking reflex is an incredibly strong instinct and if that's working OK, you can allow yourself to think you're on safe ground. Martin had put his fingers in Bambi's mouth and she'd sucked hard.

'But, Johnny,' he said, 'she is distressed. She's calling for her mum.'

'We'll come right away,' I replied.

Julie and I, along with our granddaughter Roxy,

jumped in the truck and went straight into South Molton. We walked into the surgery and we were shown through to the cages, where all sorts of animals were undergoing various treatments. And there was Bambi – I heard her voice. She was in one of the bigger cages, down at ground level.

I was really choked when I saw her. No leg – gone! And how bare the wound looked, to my eye: a 14-inch cut, right across her rump, very neatly stitched. The stump was moving involuntarily as she shifted position. It was heartbreaking. Martin Prior had warned me that when the fur is shaved off, an animal's skin looks more like a human's. It was a shock. The pad of flesh looked for all the world like a big ham. Plump and swollen. And there was just the one leg left, on the other side, which looked far too frail to do the necessary work.

She was 'meowing' and she was distressed. She was struggling in her cage, and all the time giving me a look that in the future I'd come to know well, a look that sort of said, 'Well, you're my mum, aren't you, it's a bit odd but that's what's happened, so come on, get on with it.'

I lay down in her cage with her – it was cramped but I could hold her. I stroked her head and neck. 'It's all right, Bambi,' I said, and her calling grew less.

I kept talking to her, 'It's all right, my girl, you've had an operation, and you've been under anaesthetic, Bambi, and now you've woken up, and all the poison is gone. That leg was going to kill you, Bambi, but now it's gone for ever. And you've done ever so well. You're on the mend. You're going to be all right. You'll just have to learn to run around with three legs. And you will, I know you will.' I went on like that, knowing of course she couldn't understand me, but it was the tone of the voice that was important, so I rambled on.

She grew quieter. She could tell what was happening, in the sense that she knew things were all right. If my voice was calm and kind, then it was OK.

And a minute later, she stopped calling out altogether, and became dead quiet.

It was true, I realized – I was her mum. Of course this gave me an amazing sense of how important I was to this animal. As I've said, until my own accident and injuries, I was known more for shooting deer than for saving their lives. Even as a child I'd stood at my uncle's knee and watched him take apart a shotgun cartridge, remove the shot and instead put in a single slug of lead that he'd painstakingly crafted himself, and it turned his 12-bore into an elephant gun, more or less. And then I'd seen that slug go through one deer and hit the one behind it, and kill both. I'd even found the slug and returned it to him. As a young man, when I was called up for National Service, they found they had a ready-made marksman, without having to teach me a thing. During my twenties and thirties I could more often than not be found trotting round South Molton carrying sides of venison over my shoulder through the dark, trying to avoid being spotted by my father-in-law who ran the butcher's shop, and always having to run away from the police. Or taking the poacher's wire out of my back pocket and pulling salmon out of the river by their tails.

But now, here I was, scrunched up on the floor of the vet's place, a surrogate mum to a little calf called Bambi. And she was in my arms, quiet as a mouse. She'd been soothed. She was calm. Our only thought was to get her home, just as if she were a son or a daughter.

We had the truck backed up to the vet's, and dropped the tailgate. I remember there was that same blue-checked blanket we'd put in there, for her to lie on. I carried her out, all wrapped up to keep her warm, and lifted her into the back. I climbed in there with her to keep her quiet, while Julie drove, with

Roxy next to her. Bambi didn't panic, didn't struggle. She was quiet, she was all right as long as she could feel safe, in my arms.

I can remember asking myself, as we drove the few miles back to Bishop's Nympton, the two of us bumping about in the back of the truck, did we still have a chance of getting this little deer back to the wild? That's what she really needed. The worst was over, I thought, and we should begin to work out what we'd need to do, a plan. Just the other day, Martin Prior was telling me, there was a baby otter found round here and the vet saved its life and took it to the otter sanctuary. They kept it in a big pen to protect it, but they went to a lot of trouble to make sure the baby otter never saw or heard a human being; they took great care not to allow the otter to realize it was being fed by humankind. This was to stop it from imprinting in the way that Bambi was doing, so the return to the wild would be easier. It was being cruel to be kind, for that little baby otter to have no parents, no human mum or dad. And then its transfer to its natural environment would be successful, it would find a mate of its own, a family of its own, and that would be a greater satisfaction.

However, this kind of treatment already wasn't possible with Bambi. Her injury meant she'd had to be handled by us and she'd imprinted on me. It would make it more difficult, but we'd have to keep one eye on where we wanted to go with this. My first concern would be to keep Bambi's wound clean, and clear of infection, so it could heal properly. My eye was drawn to that cut the whole time, watching for what it was doing, checking for any bleeding or whatever. Pray God it would be healed, and then the next challenge for us would be to get her fit enough. There was no doubt the missing leg would slow her down, but the question was, by how much? It would stop her jumping so high, but hopefully she'd be able to find her way through the racks and the hedges. She'd have difficulty following the herd if they were jumping in and out of farmland. We'd have to be careful to rehabilitate her with a herd that stuck more to the open moor. And of course the missing limb would probably lead to arthritis later on – Martin had warned me about that. The straightness of her spine would be thrown out of true, and the odd angle of the remaining leg would mean the bone worked in the socket just slightly wrong, but she could have a decent, full life in the wild, even if she didn't live as long as she might have done otherwise. We just had to get her on to solid food. As soon as she could forage for herself, as soon as she was in charge of her own food, if you like, which would be in four or five months or so, then we could start out on a programme for getting her back to the wild. Perhaps the first step would be to find a big enclosure, a farm, where she could be safe but have enough ground to cover, to get her up to speed. Wean her off human contact, and put her in touch with her own kind. Next, maybe we'd go up to Polworthy, and find her original herd, see what they were up to. I could see in my mind's eye how she'd wander up, a young hind, and the rest of the herd would be unsettled by the new scent, by her strange three-legged gait, but it wouldn't take long, they'd take her back. I was sure we had a chance, at least.

And then, I could imagine a bit further ahead in the future – I'd be pulling on my camouflage gear as usual, picking up my cameras and my field glasses and heading on up to the moors, every now and again, just to have a look at her and see how she was getting on, and of course take some photographs. After all, she'd be easy enough to tell apart from the others, with only three legs.

Like this, Bambi would have a shorter life, but a better one. There was no way, though, she could ever have a calf of her own. That would be something always denied to her, I thought. She wouldn't be able to stand for the stag. But no matter. She could enjoy a full life. How proud we all would be, who'd saved her. We just had to get her that far, safe and sound.

These were my idle thoughts and dreams while we drove to Bishop's Nympton after fetching her from the operation. I looked down at this tiny creature on my lap, her back covered with that scattering of white dots. It's no wonder they put baby deer on so many Christmas cards. They are magical to look at – so graceful. And yet poor Bambi had this ugly rear quarter, shaved and swollen, a great big wound stitched up, and missing that leg. It ruined her perfect good looks, and it broke your heart.

One step at a time.

When we got home, we parked out the front and carried her in. There was one occasion, many years ago, when I could have been caught skidding to a halt on this same patch of road, in front of this same house, and going round to the back of the rickety old Morris 1000 van I had in those days and hauling out an entire stag, not only dead but frozen solid, and having to move very fast with the knives and with the chainsaw, right out here in front of the house, because the carcass was thawing out, and I had to cut it up to sell for dog meat – ten quid in the back pocket, and another bit of work done.

And now it was the same man, the same Johnny Kingdom, with the same tattoos, a bit less of the curly hair, a fair few years older and some more wounds, a bit smashed up, but how very different was my behaviour with this little deer calf, now. We carried the three-legged Bambi ever so carefully inside, into the front room there, past the cabinet with the china animals in it, through the kitchen, out the back.

The garden shed was waiting for her, but we'd made a change or two while she'd been undergoing her operation. We'd got on a bit and done some work on her behalf. We'd cleared out a bigger area and made it safe and warm ready for the new arrival. Home sweet home, it should have said over that shed door. We laid her down and watched her settle. That nose was flexing away as she took in all the new smells. Paint, and garden tools, and musty old bits of wood that had been lying around in there for years. Hardly a bouquet, as far as she was concerned. But it would do for her, for the six or nine months we thought she'd be staying.

Here she was: Bambi. No more poison in her leg, making its way round her system; we'd taken care of that. No more operations. All she had to do was make herself at home in her newly tidied-up shed.

We had sorted out the goats' milk – there were two big bottles of it in the fridge and more where that came from, thanks to Jill and her herd of goats. The blue-checked blanket was there, to keep her warm. Tonight I would sleep well. The worst was over.

6 A bigger crisis

Straightaway, a worse thing happened.

It maybe wasn't such a dramatic and heart-stopping thing as the operation, but it was more dangerous. She started to scour. That is the term for what we'd describe as diarrhoea. In such a young creature it can lead very quickly to serious dehydration and then all the internal organs stop working. It's the same with human babies: the younger they are, the quicker they can lose water – much faster than you can put it in. Dehydration was the enemy.

Julie loves animals as much as I do, but I began to realize that our reactions to Bambi were different at this time. Julie began to think that all we were doing here was watching a vulnerable, young, wild animal die. And not only that, we were stringing it out; we were making it worse, a longer period of suffering. In the wild Bambi would have died quickly; it would be all over by now. Julie began to withdraw, to hold back. She didn't want this to happen in front of her eyes. She didn't like it and she couldn't easily go near Bambi. The difference was that she was convinced Bambi was going to die, whereas I stumbled on, blind to that possibility and determined that she wouldn't. Nine times out of ten Julie would have been right, and I would have been looking back and thinking it would have been kinder to have let Nature take its course, all we did was frighten her more, put her through more suffering . . .

But it wasn't in me to think that, even though it was right-minded.

I became Bambi's nursemaid. I had to keep topping her up with fluids, night and day. I spent more time in that shed than in my own house. The alarm clock – I can still hear that 'Brriinnng!' and I can still see that luminous dial with the numbers round it, one to twelve, waking me up every couple of hours. Bang the little button on the top – the alarm stops. Haul my weary bones out of bed again. Pull on the same clothes, stagger out there to the garden shed, yet again. On and on. Day after day. Dioralyte, just like for babies, to replace lost fluids and salts. The awful mess. Falling asleep in the chair in the daytime. Like an old man I'd have to sit down for a minute and the moment my head rested against the back of the chair I'd be asleep, waking up with a jerk a few minutes later. Broken sleep, when it goes on for a little while, is enough to drive you mad, but my mind was on Bambi all the time. Keeping her going.

And the wound had to be kept clean through all this. Her shed had to be spotless. The flies were a problem. This was late June, early July, the season for them. I'd walk in there, beating at them with my hands, chasing them out, swearing and cussing. Martin Prior gave me a spray to fight them off with. I couldn't bear the sight of any flies near that wound. It was enough to turn your stomach.

All these hours of nursing, the Dioralyte, the cleaning-up, the talking to Bambi – it paid off. Because Bambi survived. It was touch and go, but, slowly, she came through it, out the other side. Martin Prior had been with us all the way, giving us advice on what to do, and he should know how much I thank him, for that – from the bottom of my heart.

I could sleep, now! How about that? It's what I remembered most. A whole night. The clock just ticking away, no alarm. No more staggering out to that shed in the middle of the night, always dreading that I'd find Bambi unconscious or dead. Instead my head on the pillow all night, and then the early rise, the beautiful mornings, the birdsong, and a healthy Bambi impatient, waiting for her breakfast, calling for me. And when I went out, her eyes were bright, inquisitive. She was on her feet. Her coat was glossy and her appetite good. The kind of talking between us, the way she lifted her head to me, was a special kind of unspoken communication between us.

Only a few weeks afterwards, Martin Prior came to visit our back garden again, this time to take out Bambi's stitches. It felt like a real achievement; it meant a lot. There was a party atmosphere in our garden that day. The wound was healed! I lay down next to Bambi while Martin worked away with the scissors, snipping the threads and pulling them out, one by one. He was looking at the site of the wound, where the hair had grown over by now. The scar was almost invisible underneath. 'You've done a good job with this one, Johnny. It's clean as a whistle and it's healed well.' He smoothed his hand over her rear quarter. 'Quite a nice stump,' he said. 'And the most important thing is there's no pain. Look, she's not flinching or anything.'

I felt proud, not only for myself but for all of us

who'd kept Bambi going. Already, it was a fair old list of names, those who'd brought her on this strange journey: Merc the collie with his sensitive nose; Brian Buckingham and his quad bike; Julie and I carrying her back and forth and giving her the shed as a home; Jill and her goats providing milk; Martin Prior and his team's life-saving medical intervention; our granddaughter Roxy. I've got this lasting memory of little Roxy, only tiny she was, dressed in her pink dressing gown and slippers, taking a bottle of milk into the pen to feed Bambi. Bambi getting the hang of it and working away at that bottle. Me calling to Roxy, 'Hold 'un up, Roxy! Hold up the end!' and Roxy doing her best to handle this bottle and this deer calf that was as big as she was, to get it so the milk went downhill, down the deer's throat.

The team was only going to get bigger. Louise, my other granddaughter, who was born later than Roxy, got involved too, later on. Joe Drewer. And many others, as will become clear.

7 A deer's life

The discussion that we'd been having, off and on, ever since we found Bambi, now returned: what about her rehabilitation, releasing her back to the wild? The truth was, there had been a fair chance of doing that successfully if we could have saved her leg. There'd been a smaller chance of pulling it off when she was reduced to three legs and certainly she'd have had a much shorter life in the wild. However, the scouring, and our having to nurse her so intensively during that period, put an end to the idea. She had bonded so thoroughly with me. It was talked about a lot among all of us who cared for Bambi; we went all round the houses on this one, not least because it was such a big commitment on our part – not something you undertake lightly. But in the end none of us believed she could survive. She wouldn't have been able to stand for the stag. The hunt would have culled her – she'd have been shot. To send her back out to the herd would be as good as signing her death warrant. For better or worse she'd become one of us, a member of our family. In many ways this was a sad day for Bambi, she would never have the life she should have had.

I can remember brooding over this, during those months when we were having to come to terms with the situation. Usually when I'd had a pint or two of cider, and I'd be leaning on the kitchen counter there, looking out of the window at her shed and her fenced-in bit of garden, I'd ask myself this question: How do you give a wild animal as good a life as is possible in a back garden? This is what I wanted. This was my ambition. But how? What could I possibly do, that was in my power? Some answers might come to me, I thought, if I imagined her life as it would be in the wild, what would happen to her and when, and then we could try and give her as much of that as we could. So what is the life of a deer in the wild, I thought, hanging on to that kitchen counter, and maybe Bambi strolled out at this point, and looked in my direction, and caught my eye. Of all people, I should know how to make her life as contented as possible. I've watched the red deer of Exmoor since I was a child.

I told myself to imagine this: what if it had all gone differently, that day back in early June; say that her hind leg was never caught. Instead, she followed her mother into the field . . .

She'd have lived with her mother. Well, I was her mother. So I must be a good one, I must give her enough of my time. That was the only answer to that bit. To make a bond with her, nurture her.

In the wild, Bambi and her mother would have rejoined the herd, and together they'd have wandered from one spot to another, as a group, to find the best food.

Well, I could do that bit, too. I knew exactly what the red deer liked best to eat. I'd make sure to give her only the best.

She would have spent the summer finding her feet, working out a place for herself in the hierarchy. The games the calves play – running and jumping and play-fighting – mean they sort themselves out into groups; they make alliances, friendships. This playful part of the deer's life is marvellous to watch. Like the young of any species, they are full of life and spirit.

I'd have to make sure that Bambi, in our garden, and without her natural-born playmates, had plenty of fun and games. Animals get bored the same as humans do, and we would, as it turned out, find any number of different ways to make life fun for her. She wasn't just going to hang around watching the world go by. She'd box with an old hay bale I'd put in there for her, bashing it with her hooves like she was Mohammed Ali, literally standing up on her hind leg and cuffing that bale, giving it hell, and then running around outside her pen as fast as she could, like she was on a racetrack. 'Steady there Bambi, steady!' I'd tell her. With just the one leg propelling her along, it looked like a sure thing that she'd fall over, but no. It was amazing how she could turn on a sixpence, at speed. She'd give these big, playful leaps. Another favourite was the wrestling with me. Or she'd run in and out of the water when we turned the hose on her, in hot weather, and try and catch all the drops in her mouth, and shake herself and skitter about.

The summer is the easiest, most relaxed time of year in the wild herd, and somewhere in the pecking order there would have been a place for Bambi, it didn't really matter where. She'd have settled down. She'd have seen all the males, the stags, getting on well, living together as friends, their antlers growing more splendid daily. Then, during August, she'd have seen something strange happening, which would give the first warning of the mayhem to come. The males now become preoccupied with rubbing the velvet off those newly grown antlers. They find a tree and rub and rub against the trunk and branches, waving their heads up and down. It looks like it must be itchy, like it's making them mad. The Exmoor National Park people don't like this behaviour from the deer because it means they debark the trees, rubbing away, and that will kill the trees if it goes on for long enough. The velvet peels off, bit by bit, and drops to the ground, revealing the new antlers underneath. And, if you go and look, you can find the velvet lying strewn at the base of the trees – it makes food for the flies, who swarm all over the stuff. This is the time, also, when the stags sharpen the points on their antlers, so the trees get more punishment.

The females, meanwhile, are concerned with raising their young, keeping them close by, only weaning them after many months – and the relationship between mother and child within the group continues for years. A hind will become pregnant normally in her second year, but the family groups keep going, with mothers, daughters and granddaughters sticking close together.

Then, in October, as the days become shorter and more chilly, and the autumn mists began to wreathe the bottoms of the valleys in the early mornings, Bambi would notice a change come over her group. The calm, drifting migration of the herd, here and there, always leisurely – unless there's an obvious threat from predators – is no more. It used to be that they visited one spot or another almost thoughtlessly, or as if everyone decided together. Suddenly there's a lot of uncertainty. And – for the first time – there's aggression within the group. It's like all the males have had too much cider. The herd is anxious, volatile, almost like those crowds you see spilling out of the clubs at closing time in town centres. The herd is taken this way and that, sometimes split up. What can be going on?

The hinds have come on heat, so the rut has started.

The males can no longer be friends. Their blood is up. Now they're in competition for the females, and that makes them sworn enemies. It's the most powerful and determined stags that will win. Those with the finest antlers, the strongest physique, the most aggressive. It's natural selection – the survival of the fittest. The females can only watch, and be driven here and there by the males who are trying to win them.

Each hind is only on heat for a day or two, and so it's urgent. The males 'stand the rut', always at the same places as before. They rub their antlers against the bushes, leaving their scent to mark their territory. They roar their challenges. The deeper the roar, the more powerful the stag. Bambi, if her life had taken its natural course, would have seen the males in the group fighting, running each other out of the herd. And these fights are not just for show, they are dangerous. It can be a fight to the death. During the rut the stags will lose half their body weight. It's not that they forget to eat, exactly, but they spend all their time trotting, rounding up the hinds, moving them off, rounding them up again, standing the rut, fighting, rounding up the hinds again, moving them, and of course the whole be-all and end-all of the whole process: the mating. It's a time of frenzy, and the young males get what they can, sometimes even nipping in to steal a hind's attention while the bigger males who control the herds are fighting one another. Bambi might have found herself in a bigger herd, or a smaller one. It's a time of sound and fury, as they say. The noise of stags roaring is unmistakable.

As a matter of fact, as I write this, in October, I've agreed to judge what must be the oddest competition you can think of, if you look at it from a deer's point of view. It takes place over near Dulverton, and it's a competition to find the person who can give the most lifelike imitation of a stag's roar. A couple of hundred people will gather and make a start on the cider and the beer, waiting until it gets dark. The competitors then head off into the night. No one knows who's where, out on this rough piece of ground. Everyone strains their ears. And then it comes, out of the darkness, from different quarters, what sounds like a load of stags roaring. It's taken deadly seriously, of course, as well as being a bit of a laugh. Us judges will listen to all of them in turn, and make a decision as to which one sounds most like a real stag. We're talking about people here who for one reason or another have made it their business to imitate the rutting stags as closely as they can. And it's uncanny, how good they get. Not a little frightening for the children, either, hearing those roars come out of the dark from all around you, out in the middle of the moors. When we've made our decision, then the person steps out of the dark to reveal himself or herself and we have our winner, and then we can start to celebrate. I like to think of all the deer round about, listening to this roaring and wondering what on earth's going on ... The winner was a guy named Elvis.

So, during the rut, Bambi would have been run this way and that by the feuding males. And the rut lasts longer now, because the warmer temperatures keep it going right the way through November.

Well, there was no way I could give Bambi that experience. She might hear a stag's roar from our garden and wonder what it was all about, and she might listen to my son Craig, who does a pretty good imitation of it, but that whole process of the rut would remain unknown territory to Bambi. It was a shame. Some of the most dramatic times I've ever had, observing red deer, have been during the rut.

In December, if she were living in the wild, she'd have noticed everything begin to calm down. All that anxiety and movement would diminish and stop. The hinds are all pregnant. The rut is over. The same stags who only a week ago were fighting each other to the death are now living peacefully alongside one another, as if nothing had happened.

The females will gestate for nine months, just like humans. The task for the wild herd now is to find enough food during the cold winter months. There might be snow; there will be bitter cold and wind. They must seek shelter and keep warm. The grass has shrunk back to its winter state, grey and tasteless compared to spring, and with far less nutritional value. They will have to pick up the best they can find. They will inevitably feel hunger.

For Bambi, of course, there would be fresh hay all year round, as well as cake feed, and she would have the shelter of her little house. She wouldn't go through the same trial of winter time. These might be the months when the wild herd would look at her with envy, instead of vice versa.

In early March, with the days becoming longer and the sun rising higher in the sky and the first hint of the grass about to grow again, the whole moor emerges into spring. And with the heat of the sun on their backs, as it were, the task for the wild stags is now to help each other knock their old antlers off. It's one of the most amazing sights on Exmoor. It's like play-fighting – the same kind of action as when they were fighting for real, but without the aggression. You can tell their antlers are itching, and they spar with each other or they rub their antlers against trees until the antlers drop off, like a nut falling out of its shell.

They're elaborate constructions, antlers are. And, like teeth, you can use their size and development to determine the age of a stag. The first year, they don't grow anything at all. In their second year they'll grow two straight antlers, called prickets or uprights. In their third year they'll grow their brow points. The year after that, the bay points will grow. In their fifth year, the tray points. After that, the accepted wisdom is that they grow an extra point each year, although that isn't strictly true. It depends on what food they're eating, what their health is like, all sorts of things. But it's generally true enough to be taken as a rule of thumb. And when a stag has a brow, bay, tray and three-a-top, he's called a Royal Stag, that's to say a mature one, and that term comes from the hunting days, obviously, because royalty have always been mad keen on hunting, and the Royal Stag would be left alone by the hunt, because he is the top specimen for breeding. Once he starts going 'palm-headed', he's deteriorating; he's fourteen or fifteen years old by now, and he'll be culled.

These big sets of antlers are amazing constructions, especially when you take into account they've been grown in just a few months, and it's a hell of a scene, come early March, the amount of people out there hoping to find a good pair. I'd guess there are hundreds involved every season. Bambi, if she lived in a wild herd at this time of year, might notice the torches flashing over the ground in the early hours, as people search – maybe they've got one antler, but with the other one as well, the pair will be worth a couple of hundred pounds. And that second antler might be twenty yards from the first, or three hundred yards, or even a mile away. They scour the ground, hoping to make the early find. It's a fairly intense time and there's a lot of luck involved, and a fair amount of skill. If you find one good one, it's a question of tracking that deer and finding the other one – no easy matter when there are others chasing after the same thing. Money changes hands and you can imagine what a difficult bargain it is to strike – if you've got one antler and the other fellow has the other one, what's it worth to give yours up, or what will you offer to buy him out? My cousin Terry Moule sets antlers in a plaster mould. It's important to set them 4 inches apart, so they sit right on the wall. There's a fellow called Tom Lock, from up at Hawkridge, who's found a lot of antlers in his time, and he's famous for making things out of them: candlesticks, whistles, walking sticks, riding crops. He even made an armchair out of antlers. I'd like to know what that cost.

As soon as the stags have lost their old antlers, they start growing new ones. And for the next few months, all the energy of the males, every blade of grass they eat, goes straight to the head. It's an important job: they have to grow a new, bigger pair of antlers from scratch every year. It's a lot of work. It's as if you can watch the sweat dripping from their foreheads, they grow them so quick.

Meanwhile the hinds, come the month of June, approach the time for labour and birth. They will separate off from the herd, find a quiet spot . . . and have their young. And this was something we definitely couldn't do for Bambi. We'd never be able to breed from her, even if it had been possible with artificial insemination. It's one thing to rescue an injured creature, who otherwise would have died, and give her a good life. It's quite another thing to decide to bring a wild creature into this world deliberately, for the sake of the mother. That wouldn't have been fair on the calf. So there was no way Bambi would ever have a baby of her own. 'Sorry, Bambi,' I'd say, scratching behind her ear. She had this habit of lifting her nose high in the air; it was one of the ways she talked to me. 'That's not for you. Bad luck, my girl.'

8 A marriage for three

Nevertheless, we wanted to give Bambi as many of the pleasures she'd have had in the wild, if we could. My mind turned to the idea of giving her a bit of freedom to wander, and then she'd see new sights, and enjoy new smells. I believed I could do this for her. I could put a collar on her and train her to be led on the end of a rope, like a pony, and if I could do that, then I would take her out of the garden and round the front of the house, and if that went OK, then perhaps I could take her to different places.

I first put the collar on her in the garden, in the safety of the pen, and she didn't like it one bit. I quickly realized I'd made a mistake. I should have put a collar on her from day one, and then she'd have grown used to it from the start. It would have been a different story, then. As it was, she had a few questions about it, shall we say.

'It's all right, Bambi,' I said, 'come on now. Mind your manners.'

It seemed like she didn't want to get the hang of it. I wasn't terribly optimistic. It felt all wrong.

The next challenge was to walk her out of her enclosure, around the side of the house and out the front, ready for our small adventure. I wanted it to work; after all, if it did, there was no telling what could be done. The world would be our oyster.

Bambi had the notion in her head that if she was outside her enclosure, then it was because it was time for her to go racing round the garden. And it's true, she had made a racetrack for herself around the very outer limits of the garden and it was her pleasure to show us how fast she could go round it on just three legs. It was phenomenal to watch. I could swear she took real pleasure in us clapping and hollering. I always ended up being worried she'd hurt herself and I'd cry out, 'Steady, Bambi, slow down!' However, she needed to understand that when she had the collar on, a different set of manners was required. It would be no good if she just took off. I'd end up being dragged through the mud, face down. Suffice to say that it took a while for her to get the hang of the new arrangement.

Eventually, though, it was time to show Bambi that there was another world, around the front of the house. We walked together down the very narrow passageway between my house and the neighbour's fence. She was hyper alert, you might say. I was telling her in my calmest voice that everything was all right, that she was safe, that she must behave herself, for all our sakes.

When she came out the front, all hell was let loose. It was as if she said, 'Hey, you didn't tell me about this, look, and look at that . . .' She headed off in every direction at once, dragging me along with her.

It wasn't only that she was excited, I could tell from her body language that she was also afraid. I dug in my heels and hung on to the rope. My ears burned. I could imagine my neighbours watching and saying, 'Look, there goes Johnny Kingdom, he thinks his red deer is a pony.' It's hardly a dignified thing, to be dragged around by a three-legged deer.

'Bambi!' I called. 'Settle down now. What you doing? Where you going? Behave yourself. This isn't any good . . .'

She didn't take a blind bit of notice.

If only she'd have walked with me, like I thought she might of! There'd have been no end of places we could have gone together. It would have added to her life, no doubt about it. But as soon as she got out the front of the house, it was impossible to hold her. And I dared not let her go. She'd have run off, and I couldn't be doing with that. The mind filled with terrible possibilities. A collision with a car, and Bambi either lying injured, or causing injury to someone, or both. Or maybe she'd become lost and wouldn't be able to fend for herself – and that would be a slow and miserable death. Most likely, I'd receive a phone call from a neighbour more or less far away to say, 'Johnny, your deer is here causing a nuisance.'

So, after I'd tried and failed, I had to come to the conclusion that it was for the best if her world remained a small one. Our garden became her home. Its borders were the limit of her experience, and it was our duty to give her as rich a life as we could. The games, we could do them all right. The human companionship in place of the companionship of her own kind. The food – I made big efforts to make her work for it, hanging the choicest bits in different parts of her pen so she'd have to browse along a bit, like they do in the wild.

And, looking back on her years with us, I can only think she had a wonderful time, in many different ways, even with such limitations on her freedom.

When we accepted this fact – that she was with us for the course of her natural life – we had to build a bigger pen for her. As it turned out, it would get bigger and bigger, in stages, at the same time as she did. Three times it was extended, until it took up the whole garden. It was my sons Craig and Stuart who did the work. It was important to give her as much space as we could.

It meant that Julie's garden suffered mightily, of course, which added to the triangular relationship that was developing between Bambi, Julie and myself. As Julie says, it's like Bambi thought I was her property, and Julie wasn't allowed near me. She put on a haughty air when Julie was there. She took the opportunity, every now and again, to give Julie a little nip or a push. Bambi has always been jealous and I believe, I do, true, that it actually annoyed her to see that Julie was able to talk to me, and that we could put our arms round one another, and we could sit together taking a cup of coffee in the morning, and, worst of all, Julie appeared to be living in the house with me, whereas for some reason Bambi was out in the garden. That really stuck in her throat. She flicked her ears back and forth and you could see her thinking: Why aren't I in there? What's she got that I haven't . . . Really, it could easily have got like that children's story book, where the animals come in the house and take over. If Bambi could have turfed Julie out the house and taken her place then she would have done so at the drop of a hat. I'm not sure Bambi would even have been so generous as to keep Julie in the shed, if the boot had been on the other foot.

So it seemed like part of Bambi's whole revenge thing going on, her punishment of Julie, that she ate all of Julie's very best roses. I don't think my behaviour helped very much either. Julie was working at the village shop at this time, and I'd always wait until she'd gone to work before I let Bambi out to have a free run around the garden. And then Julie would come back from the shop and every now and again she'd bring Bambi back an out-of-date doughnut, or some treat like that, and Bambi would say thank you very much and eat it, with the jam dripping down her chin or whatever, and then, as her reward for this act of kindness, Julie would find out that more of her garden had been eaten. I was caught in the middle, of course. If I wasn't getting these silent complaints from Bambi about Julie's existence, then I was having to cope with Julie having lost her roses. It became like one of those marriages with three people in them, 'a bit crowded' as Princess Diana said that time.

The truth was, Bambi reached over the fence and ate anything she could. The fence grew bigger – and so more of the garden disappeared.

And she had her own home, of course. The shed was completely cleaned out, except for that blue-checked blanket that was always to hand, and all the stuff was taken to Stuart or Craig's house, or put elsewhere. She was weaned off the goats' milk and on to hay and grass cuttings and Coarse Calf 18 – a cake made up of cereals and molasses and protein. She loved that. It was Adrian, my neighbour, who started giving her his grass cuttings. I imagine it was quite useful for him: he could get rid of his garden waste. He was a farm worker, so he brought back hay for her as well, and also looked after her when we were away. Another neighbour, Mike, one over from Adrian, had a good line of jokes at Bambi's expense. If he was having a barbecue he'd peer over the fence, with his apron on and his barbecue fork in his hand, and when the charcoal was up to speed he'd call out, 'Right, Bambi, we're ready for you. Let's have that other leg now, please.'

As well as the cake and the hay and grass, I'd

wander along the lanes for half an hour in the morning and the evening, cutting out young ash shoots and hazel leaves. Honestly, they could have paid me as a hedge-trimmer round our way. I'd bring the shoots and the branches back home and tie them all along the fence, so Bambi could browse along and pick what she wanted. I tried to make it look as wild as possible, so she had to do some work, forage for herself.

There were others who helped with her feeding. And I can't even begin to say that without straight-away mentioning Joe Drewer; he had such a lot to do with Bambi. He was a wonderful bloke and a close friend, and, luckily for Bambi, he ran a very good vegetable garden and he lived only a couple of hundred yards up the hill from us.

He worked for the council, did Joe, clearing the ditches and waterways alongside the country lanes. So he'd put on his overalls and they had the small truck they drove in, and they went to the particular lane that needed doing, and they'd work all day with proper Devon shovels, clearing and digging and making good the drains and ditches. And I have a lot of memories of Joe Drewer shovelling snow over the years as well, during the thickest parts of the winters, and that's partly because I'd sometimes work with him. Joe and I have moved mountains of snow in our time. Then, when he got home, he'd work on his vegetable garden. This was his pride and joy. In this way, he was like my own father, because he won so many prizes in the vegetable and flower show in Bishop's Nympton. His house, just up the road from my place, was always surrounded by a riot of colourful flowers, as well as his vegetable patch and his greenhouse, of course. It was always immaculately tended, very colourful – and remains so to this day. He had what I will describe as his own private temper, did Joe. He was a peaceful man and never caused anyone a moment's trouble, but if someone's dog wandered into his garden and started scratching at a flowerbed, mind out. I could hear him from all the way down here. 'Get home, you!' he'd shout. 'Sod you!' There wasn't a soul that would cross Joe Drewer.

And then, when he'd finished his work for the council, and after he'd finished his garden and his greenhouse, it was time to knock off. He'd go indoors and he'd change out of his overalls into smart, clean, neat clothes – and it was never anything less than a collar and tie for Joe. That same kind of precision and tidiness with which he did his garden, and worked alongside the roads, he attached also to his physical appearance and to the inside of his house. Everything was orderly and in its proper place and well maintained. He had a thin face, and he was tall and elegant-looking, and a calm man. That private temper of his was locked away, almost all the time. And when it came to Bambi, there wasn't a more gentle or generous person.

Very often, in the morning or evening, either side of his other duties, he'd stroll down the road – quite a steep hill, we're on – with a few bits and pieces from his garden for Bambi. It was Angela, his daughter, who told me the other day that it was always the crooked vegetables or the small ones that Bambi got. It was because he was such a champion grower. He'd sort through his crop and if there was one that was growing a bit bent, or if it was a bit of a runt, out it would come and be brought down for Bambi. She'd call him, 'Mrrrrr!' and come to say hullo, see what he had in his pockets, and in that bag he carried. And this went on for years. He got genuine pleasure, did Joe, from his relationship with Bambi. He looked after her as much as I did, and he was like a favourite uncle to her. When we went away it was Joe Drewer who'd help out, visiting Bambi as much as was necessary, managing her food and water and exercise and medical care. And he helped me a great deal in other ways, too. We ended up travelling large parts of the country together, Joe and I, selling my videos and photographs at country fairs.

And, of the most recent graves that I've dug, the one that is most pertinent to this story, the story of Bambi, is Joe Drewer's. There wasn't a kinder man than he was. I remember wandering up there so often, and he had a blackbird in his hedge that I filmed, and a rare song thrush, and always a host of beautiful butterflies and moths, because he had such a palace for a garden. He was a dear, dear friend of mine, and of Bambi's. He looked so peaceful in death. A suit and tie he wore for his grave, and his silver hair as well groomed as it ever was. He was buried here at Bishop's Nympton, and although I dug his grave I didn't backfill it, seeing as I had to change and attend the service. My boys filled him in for us.

When he died, I swear Bambi noticed. Where was that man who always came with the crooked vegetables, who scratched her back with a broom, who talked to her like I did?

After Joe died, his wife Rene and their daughter, Ang, carried on bringing down food for her.

And our neighbours, Linda and her partner Justin, were brilliant at looking after Bambi when we were away. Linda remembers a time once when there happened to be thunder and lightning on fireworks night. Bambi never liked thunder and lightning, and she never liked fireworks, and this night she got both at the same time. She was scatting about like a mad thing, frightened half to death. Linda and Justin had to call Craig and Craig and Justin got in the pen with her to calm her down. They had to leave the outside light on. Linda was brilliant, and since she worked in the fruit shop she brought back treats for Bambi – strawberries and all sorts. Linda showed Bambi a peach, once, and Bambi just took it and ate the whole lot, stone and all. That scared me; I hoped the stone would come out, but it must have done because she was all right.

Our cat, Smoky, also became a good friend of Bambi's. Smoky was black and white with four white socks. She was used to sharing her garden with all sorts of creatures, was Smoky: four lambs, ducks, chickens, loads of different animals that I'd brought back for one reason or another, from the farm up at Sindercombe – where she'd come from, herself – but she was a bit mystified at this new one, this creature that was growing so big. Smoky would run around the enclosure, with Bambi chasing after her. Or she'd sit on one of the posts and Bambi would come up and put her nose up to kiss her. They became mates, as animals will, who live together. Often you'd find them lying down together. Dear old Smoky, who was famous for never visiting the vet – she'd rather claw her way to Australia than leave this house. She's buried in the garden. Bambi outlived her; you'd never have thought it.

And of course people from the surrounding area began to hear about this deer that was being kept in someone's back garden. And the visitors started coming – and never really stopped. Strangers, that is, members of the public, not just our own group of friends and neighbours. They were like her official visitors. Julie remembers that the first of these was the local playgroup – they were called the Little Owls – who were going on a nature walk, and Julie invited them to come and see Bambi. The children all lined up to touch her and there was that thing that always happens with children and animals – they still have that sense of amazement at the very existence of animals, the magic of them. So they were enchanted. And Bambi herself got incredibly lively, with these children around. Her nose wrinkled and she gave her 'Mrrrrr!' and she paraded up and down the fence, showing off her repertoire of kicks and sprints and jumps, much to their delight. She never lost interest in her visitors and they became a big part of her life. She was brought gifts. In a way, she was a bit like royalty, was Bambi.

9 Making films

While Julie worked in the village shop at this time, I had several different jobs. I worked as a gravedigger, as I had done all my life, since I started helping Father with it as a child. That was backbreaking; it put a sweat on you all right, and built your strength. And I've always thought it gave me a particular outlook on life that I wouldn't have had if I hadn't dug so many graves. To be so continually reminded of death, week in and week out, makes you aware of the other side, and it also gives you an idea of the value, the preciousness, of being alive. And how quick you can find yourself going from one to the other, not to put too fine a point on it. It's wafer thin, the border between life and death, and it's invisible, and you never know where it is, how fast it's coming towards you. All you can say for certain is that it's close by, and it knows you're there.

As well as being a gravedigger, I was also a farm-worker up at Sindercombe, a moorland farm run by Herbert Thorne. It's where Smoky had come from, in fact. I loved that job. Partly it was because I grew very fond of Herbert Thorne, a wonderful man. Not only that, I was working with animals and Herbert kept his animals in beautiful condition; there was a lot of satisfaction in farming so well. Add to this the fact that you're working up on Exmoor, with its beauty coming at you from every direction, in all seasons, spring to winter, and you can see why I felt good about that job.

The other type of work I was up to, the work that was more relevant to Bambi and her life, was the wildlife photography and film-making, which came about in a very particular way. It was lucky that I had latched on to it, because I don't know what I'd have done if I hadn't. Wildlife photography rescued me from a broken nerve.

I mentioned earlier that I'd had a bad accident which nearly cost me my life, and after that accident I went seriously downhill. I'd had loads of accidents before, of one sort or another, but this one was of a different order, and it was a surprise even to me how it affected me. I didn't think I'd ever be the sort of person to be hit by depression, to find myself in a hole, have a breakdown, whatever you want to call it. Well, I only recovered thanks to two things. The first was Revd Pennington and the help he gave me, as I've described. He turned me round. But then what kept me climbing, got me out of the hole, was the second thing: I swapped the rifle and the shotgun for a very different type of shooting – with a camera. I've always liked cameras. I had a little Brownie camera as a schoolboy and I'd been very keen to take pictures of a cow, or a sheep. And it's certainly true to say that at school Art was the only subject I was any good at, apart from getting into trouble and then running away across the nearest hillside.

So when the Revd Pennington got me out of the doldrums and I was looking for a way to get back into gear, I knew it wasn't any good picking up where I'd left off. I didn't want to shoot anything, or hurt anything, ever again. But more than ever I needed to be close to animals. That's been true my whole life, from when I picked snails out from under stones when I was two, to keeping my hundreds of mice in the outhouse, rearing little baby owls that had fallen out of their nest and so on. In particular, I'd always loved being up close to Exmoor's red deer – and when I say close, I mean right up close. As a poacher

I'd developed all the skills necessary: patience, the art of camouflage, intimate knowledge of their environment and where they went and how they behaved at different times of year. The ability to track them from their footprints and other signs they leave behind. Stealth – to move quietly and invisibly, so you can get up close. But I couldn't think of firing a rifle. I didn't want to hurt a creature ever again. So what was the point of stalking them? And then I had an idea: to use a camera. The camera would give me an excuse to use those same skills. It was a new kind of shooting. Instead of pulling the trigger, I'd push the 'record' button. I knew that a friend of mine, Roger Gregory, had recently bought one of the early video cameras. A monster, it was, compared to the little palm-sized ones you get today. It's amazing how quickly the idea of tape going round and round, from one spool to another, becomes so old-fashioned. But that's how they were in those days, big clumsy boxes with spools of magnetic tape. Roger has been a good friend of mine for God knows how long and he very kindly lent me this valuable piece of equipment.

Let's just say I started off in a small way, and took small steps for years and years. The first footage I brought back was no good at all. I'd got very close to a large herd of deer, Bambi's forefathers among them, perhaps. And I'd pulled the mask down over my face, dropped to a crawl. For the first time it wasn't a rifle sight I trained on them, but the lens of the camera. I'd pushed the record button, very hopeful that these first shots of mine would be good. I felt I was back in my own skin; I was pleased with myself – that is, until I got home and discovered I'd had the camera switched on while I was walking, and so when I thought I'd been recording I'd actually switched the camera off. I had nothing but boulders and heather and a bit of the river as I waded through. Disaster, that was! A backward step you might say.

I was a pretty good shot with a rifle in my hands, the army had classed me as a marksman on the day I joined up for National Service, but I was all fingers and thumbs with the camera. I was used to looking down a barrel and having just two simple tasks: aim true, and squeeze the trigger. With the camera I had to deal with light, aperture, shutter speed, different types of film stock or magnetic tape, and hundreds of different buttons doing a hundred different things. But I didn't let this first disastrous outing with the camera put me off. So what? Everyone had to learn, I told myself. And, sure enough, after I'd been out once or twice more, I was bitten by the camera bug. In fact, when Bambi came along some years later, I was still practising, and I used to use her to practise on, framing shots while hidden behind trees, that kind of thing. Right from the word go, I wanted to get a handle on using cameras really well. It became the answer to everything. I couldn't go on borrowing Roger's so pretty soon I bought my own. I didn't have any money at all at that stage, I was just coming out of a period of being rock bottom, on sickness benefit. But with the insurance money I got after the accident, I could buy my own camera, start filming on my own account. And as soon as I started making these films, my spirits lifted. I'd found a way of climbing out of the hole I was in. And, as always, it was to do with animals, and in particular Bambi's species, the famous red deer of Exmoor.

And to begin with I didn't even think of showing what I'd done to anyone else apart from Julie. I was doing it for myself, to mend myself. And there was nothing better for my morale than tracking the red deer's footprints, and getting up close to them. I knew the wild herds like you know your favourite soap opera on TV. All the ups and downs. I filmed one stag for eight years. He became famous, and was given the name Bruno, and he had a nickname as well – 'the King of Rackenford'. He was magnificent. He once had me hiding up a tree for three hours. It was during the rut, a dangerous time, and I'd got too close to him, and he'd come at me through the bushes. Suddenly he was there, arriving pretty fast, and I had no choice but to take myself and my camera up this tree in order to keep out of his way. He deliberately kept me up there. I didn't dare climb down, not with the King of Rackenford pacing around, waiting for me. His head would be described as a brow, bay, tray and seven- and six-a-top – truly a big, powerful creature. I went back another time that year and followed him again, and he went all the way round the next farm, across the North Devon Link Road, right back round again, a big circle, and there I filmed the biggest fight I ever saw. It was like the ground was shaking. The snorting, the noise, the crashing of antlers, the trundling of hooves over the ground – it was electrifying! Eventually the other stag stepped sideways, groaning, and staggered off. The King of Rackenford had won again. I often wonder which of the stags was father to Bambi. It was unlikely to have been Bruno, but it could have been any number of the other stags that I was watching, and following, at that time.

But to begin with, my filming was very amateur. I cut together my tapes just using the pause button on the double video player. I just went around setting up my TV and player in draughty village halls, where there were a lot of technical mishaps, a lot of cursing at bits of cable and so on, and a very patient audience while I shone a torch in the back of various bits of equipment trying to get it to work. It was more like what you'd get if you were invited round to someone's home.

But, inch by inch, I got better at it. I bought better equipment, and I developed a skill with the camera, and with editing, and music. I bought other bits of equipment. Most importantly, a 100-foot remote lead so that I could stand some way back from the camera, wait, and then turn on the camera when I could see something was happening. I remember the first material I got with this was a pair of long-tailed tits building their nest. They didn't know I was there. It was wonderful. I got every bit of it, with the camera on its tripod covered in bits of gorse and leaves as camouflage, and right up close, while I stood a hundred feet away with field glasses and a little button to push on the end of a cable.

I roped in Stuart, my eldest son, to film me in the river one summer. I started off tickling trout – quite a nice one of 4 pounds, I got, with Stuart filming away. Then I saw a salmon and I took the fox-wire out of my pocket, and Stuart filmed me as I crept up behind this salmon and slipped the wire round his tail and pulled him out. I was so good at that – I'd had a lifetime's practice.

I took the camera with me when I went grave-digging. As I cut down through the earth with the spade, hour after hour, I'd glance up and perhaps I'd notice where the blackbirds were in the ivy, or where the thrushes' nests might be. Then I'd break off for a rest, and climb out of the grave, and maybe instead of sitting up against a tree and taking a drink, I'd just wander off for a minute or two, and set up the camera where I saw that blackbird disappear into the ivy, or I'd train the camera on where that nuthatch came out of a tree, and I'd get more shots. Never, mind you, would I film when there was eggs in the nest, or when the chicks were very young. I didn't want to disturb the mother at that crucial time. It's my philosophy never to disturb or alter things, just to get the shots as things happen, without any impact on the surroundings. So I'd film the nests when the chicks were nearly grown. I'd go home tired, muscles aching, but along with my shovels and stone fork and pickaxe and so on I'd take back the camera, tripod and remote lead, with a few more feet of tape used up.

When I worked up at the farm at Sindercombe, I

took the camera with me as well. The farmer, Herbert Thorne, once told me he changed his mind about me after just such a time. He used to think I was a wild man – he'd heard so much about me, that I was always into drinking and fighting and carrying on, and poaching and shooting deer and breaking the law and crashing cars and all sorts. None of that was wrong. It's true I did all those things in spades. But then he saw how I worked with his animals, how much pleasure I took in their wellbeing and how it was my natural impulse to see them right, before anything else, and he watched me standing dead still for a whole hour, waiting to get a shot of a blue tit in the hedge. And he saw how much interest I took in red deer, how much I knew about them and cared about what happened to the various different herds. He knew then that despite all he'd heard, I wasn't quite the man he'd thought I was. I'd changed.

But I could say that I was further ahead than most wildlife cameramen when they start out, because as I've said I already had the most important set of skills they need. It was the same ones I'd always had, ever since I was a child: the ability to get close to animals, and the patience to wait until you get to see the amazing thing, the important bit. I'd developed these skills while carrying a rifle or a shotgun or a piece of fox-wire. Just like with poaching – hunting of any kind – to get close to any wild animal requires camouflage, silence, stillness, and making sure no scent reaches them. I'd been doing this all my life; it was second nature to me. I could get good at this wildlife photography business, I realized. I loved doing it.

I trained myself in how to produce a good sound-track. I got used to talking to the camera while I was filming, so there was a narration, from on the spot as it were, rather than recorded afterwards. It sounded much more exciting, because I was often excited myself at what I was seeing happen, and I thought it was better that way. It suited my style, as an enthusiast who was doing this for himself, really, but wanting to share it with as many people as possible. It was pretty raw and unsophisticated but I found out, through showing my films in village halls, that more and more people were becoming interested.

I bought a classier camera, and started to sell my films in other places – I took stalls at country shows, and in country markets. It became part of the way I made my living. Little did I know it then, but, years ahead, all this would lead to a little deer calf called Bambi becoming quite famous.

Because in particular, of course, I was filming red

deer – Bambi's relatives – up there on Exmoor, on the territory I'd known every nook and cranny of since my earliest childhood, following in the footsteps of my father, and his father before him. I was willing to spend hours and hours to get a good shot of red deer. The more unusual the better.

I started to become known for it, in a small way. I remember I was at a club called Jethro's, at Lewdown near Okehampton. Jethro himself is a famous comedian hereabouts, the only one to be invited twice on to *The Des O'Connor Show*. But he is blue. There's a famous out-take of him and Jim Davidson, and they can't stop laughing. It goes on for what seems like hours. And he was doing his act in the club he runs down at Lewdown and he saw me in the audience and he picked me out, and we went up to his lounge afterwards to have a whisky, and it turned out he'd bought some of my films and he was a fan. He loved his horses, did Jethro, so we could talk about animals for ever, both of us.

And there was one particularly good sequence I'd shot of red deer, that I was so proud of. Up near White Chapple, in a lovely spot in a wooded valley, I'd been lucky enough to find a deer wallow under an oak tree. Deer like to give themselves a mud bath. It's common practice among any number of creatures and, believe it or not, it's because the mud cleans them. It's sticky as hell, and they get coated in it, all over, and then mud dries and falls off in a fine dust, taking the natural grease and oil on their fur with it, as well as stopping ticks and mites to a certain extent. You'll see lots of animals using dust or mud to clean themselves. Chickens, for instance, love to make a dry, dusty bath, and kick the dust up in among their feathers. Anyway, it was obvious that some of Bambi's ancestors went to this spot to wallow in the mud. I wanted to try and film this happening, close up. Deer are such flighty creatures, one scent or sound will set them off. Bambi was the same – you wouldn't believe how fast she could scat round our garden. And because they are particularly vulnerable when they are wallowing, they'd be making double sure they were safe. I'd have to be well hidden to catch them at it.

For four days on the trot I went up there early in the morning and waited, and waited. I was camouflaged and downwind of the wallow, but I saw nothing. Somehow the deer were getting wind of me and staying away. I'd have to think of a different tactic.

On the fifth day, I went even earlier, when it was still dark, around four in the morning. I took no torch, but made my way through the pitch black, my camera and some sandwiches and a flask in a rucksack as usual. An Exmoor wood in the middle of the night, in the dark, is a spooky place to be. I found my way to the wallow and this time I climbed into the branches of the oak tree, right above the wallow. There I settled down to wait. I heard a tawny owl calling – which scared me. I also heard a stag roar, not very far off. For four hours I sat on that branch, leaning against the trunk. I grew very stiff and numb.

Inch by inch, it grew light. I held my breath. If it was going to happen at all, it would be now.

At around eight-thirty in the morning twelve deer – hinds and yearlings – quietly picked their way through the woodland, walking with such perfect grace and quietness as took your breath away. A hind stepped into the wallow right beneath where I was sitting and daintily sat down in the mud. She lay right out on her side, then rolled on her back and kicked her legs in the air. From above, I filmed her – I was almost shaking with excitement. She rolled from side to side, over and over. Then she picked herself up, shook the mud off her coat, and stepped out.

Then it was the turn of the next one. She stepped into the wallow, sank to her knees, and rolled on to her side. Likewise she kicked her legs up, rolled back and forth. It was an incredible sight.

A little calf came next, just about the same age as Bambi was when she'd got herself caught up in that fence. I could look down on her back and count the white spots. This little one had learned off her mum what to do: down she went and rolled in the mud. Years later, when Bambi had arrived with us, I remembered this sight when I was thinking about what sort of life Bambi ought to have. Bambi did like to roll in the mud of course, but she could only do it when it was wet and the rain ran down off the high part of the garden – which got higher with all the sawdust we put down for her – and so it formed a little pond that she liked to roll in. In summertime she took dustbaths, and of course we groomed her with a pony brush, and kept her feet trim, and so on. But all that pampering and grooming that we did for her could never be the same as what I saw and filmed that early morning: a proper bathing spot to visit, along with your whole family.

I filmed all twelve of that herd of deer as, one by one, they used the wallow. Even the little calves – they all wallowed. I was over the moon.

Last of all, the stag came. He was a beauty. This was breathtaking: a Royal Stag, brow, bay, tray and three-a-top. The Queen's deer. This naming of the deer as 'Royal' comes from the times when Exmoor was what they called a 'Royal Forest' – one of many huge areas of the country that were set aside for use by the monarch for hunting, and, as I've described, it was forbidden for anyone else to use them.

This stag had obviously been keeping a lookout – and now it was his turn. He was extremely suspicious. He stood right under my tree and scented the ground. He was uneasy, but he couldn't quite put his finger on what was wrong. He looked in all directions; he tested the air. But he couldn't find anything. So he decided to take his bath. He rolled in the wallow, just as his herd had done.

When he stood up, he shook off the mud and waited. There was still something bothering him, you could see him trying to work it out.

Then he looked up. Through the lens of the camera I met his eye. He was looking at a man, his sworn enemy, his predator, right above him.

The instant he saw me, he bolted – and his herd with him. There was no more than a couple of seconds of the sound of their crashing through undergrowth, and then silence, as if nothing had happened.

It was the best footage I ever got of Exmoor's red

deer, and I included it in a film called *Rutting Stags of Exmoor*.

So, you can see it was the most natural thing in the world for me to start filming Bambi. There was no problem with access. I didn't have to wait up a tree for half the night to film Bambi. I didn't have to wear camouflage or creep around. I didn't have to worry about which direction the wind was coming from. She was right there in the back garden, and the video camera became part and parcel of the fun and games we had, the ups and downs. In fact she grew to like the microphone too much. I think it was because it wore a fluffy coat, which reduces wind noise. She had a habit of trying to get hold of that in her teeth and pulling it to bits. She thought it was a plaything.

I filmed Bambi, but not in a conscious way; I didn't think of it like work. It was just a home movie. There was only one way I ever thought of Bambi as being part of my 'work' as a wildlife photographer, and that was when I'd practise on her. I'd hide behind one of the trees in the garden, wearing a particular piece of camouflage, say a new mask or something, and I'd try and creep up on her without her seeing me. Or if I had a new piece of equipment, I'd try it out on her first. She was like my guinea pig.

But, as it turned out, people were so interested in Bambi's story that she would become a star of the small screen herself.

10 Unwelcome guests

There was a time when Bambi nearly had a roe deer as a companion, although it's not on any film, and I didn't talk about it in my last book, because it's a sad story, especially when you put it alongside Bambi's. Nature doesn't know about fairness, it's not a word in her dictionary.

Anyway, a fellow I knew, Steve – a rat-catcher by trade – called me up one day to say that he'd been driving up through the Exe valley and he'd found this little roe deer fawn, only a few days old, lying by the side of the road, trembling and traumatized.

The roe deer is different from the red deer in a number of ways. They're much smaller, for a start, and they show the little patches of white on their backsides when they run away. Their whole cycle is different. Their rutting season is in July; they have their babies in May. They drop their antlers in November. They are not so keen on each other's company as red deer – they're more solitary. I don't know as much about roe deer as I do about the Exmoor red deer. I can walk into a wood and just from the scent I can tell you if there's a herd of red deer in there, or how long ago they left. I couldn't do that with roe deer. They're much harder to stalk, being more nervous. And it's not a commonly known fact that nine out of ten times, roe deer have twins, and it happens sometimes that one of the babies gets left behind, and plainly this is what had happened in this case. She was such a pretty little thing, her spots all in rows instead of sprinkled over, like on red deer. And roe deer have the darker-coloured mouth, and the little scent pads just above the hock on the rear legs. She was still shaking with fright when he lifted her out of the car for us to have a look at. She was scratched and knocked about a bit, and I suppose she might have been struck by a car, but there wasn't any obvious damage to her, unlike in Bambi's case. I thought: Here we go again . . . but I also asked myself: Can this one be a bit of company for Bambi, for a while? Is there enough room in the garden for both of them? They'd have to get on well enough, share the enclosure, and the shed . . . but I thought it might be just what was needed, the missing element in Bambi's life: the company of her own kind. Even if it was only for six months or so, because of course we'd be hoping to return this little one to the wild, if we could. Let's see what happens, I thought. One step at a time. But, on the face of it, this calf stood a better chance than Bambi of staying alive. I got my goats' milk from down near Bish Mill, this time, because Jill had moved on to keeping cows instead of goats, and I built a little pen for her right alongside Bambi's. I didn't want to put her in with Bambi straightaway in case Bambi frightened her. It's true to say that certain animals can die of fright, they really can, and for a little baby roe deer to have a fully grown red deer going for her could have been the end, given how shocked this new arrival was already.

Bambi was disconcerted when she first met the roe deer fawn. Her nose wrinkled and she paced around, staring. She put on her haughty look, nose in the air, a curled lip, and took a turn or two around the enclosure trying to pretend nothing important was happening. But she couldn't help flicking her ears forward and staring. She then walked kind of sideways up to the fence, as if it was an accident that she happened to find herself there, and she took a closer look. The instant the fawn twitched its tail, Bambi leaped back as if someone had fired a gun.

She didn't like it. I could tell from the complaining tone in her voice – 'Mrrrrr!' She wasn't making any bones about it; she was saying to me: This is your fault, what on earth are you doing, where did you find it, get this thing out of here, this is my garden.

I watched it happen several times. Bambi would approach the newcomer, wearing an expression of mistrust, nose in the air, superior as anything, and she'd put her nose down to enquire a bit more closely what on earth this was, then she'd spring backwards as if bitten and skitter away, looking as if she were appalled at such a lowering of standards, such an intrusion in her home. The whole thing was a display of jealousy. She was very used to being the boss of her small world, was Bambi, and she didn't like the look of anything that might knock her off her perch. And to be honest, I think she took the credit for anything good that happened in the garden, while she blamed me for anything that went wrong. That's how she made me feel, anyway.

'Never mind,' I said to her. 'You'll get over it. Come on. Think about all the fun and games you'll have. It's all very well just having me to muck around with, but often I'm not here, am I? And so now when I'm gone you can have fun and games with this little one, can't you?'

'Mrrrrr . . .'

It would have been all right. I'm sure in time they'd have become friends; but time was not a luxury we were allowed, this time, with this little roe deer. Because the new arrival began to scour, same as Bambi did. It brought back all the memories of those awful days and nights, nursing Bambi through the same thing. But this time I was more confident we could get through it. After all, we had a lot of experience and we'd been successful before. And so it was all hands on deck again. The same alarm clock going 'Brriinnng!', the same broken nights, the same old staggering downstairs and out to the garden to see what had happened, always feeling that slight dread that something would have gone wrong. The endless cleaning up, the Dioralyte, the falling asleep in the armchair in the daytime to try and catch up on lost sleep. But, this time around, I woke up one morning, bashed the alarm clock, trudged downstairs, and went straight out into the garden, and that sense of dread I had, at what I was going to find . . . it was proved right. I found exactly that – yes, what I'd always expected with Bambi more than with this little roe deer. I found the poor baby creature lying dead, cold. I buried her myself, and wondered at the cruelty of life that keeps with some, and not others, with no rhyme nor reason. It made me realize how lucky we were to have saved Bambi – the injury, the operation, and the scouring on top of that. Bambi had a very strong will to live. You can feel that in a creature. Some of them quickly give up; others struggle and fight back and 'never say die'. It's the same with all of us, I suppose. I think I'm probably one of the ones that kick and struggle and agitate on the surface of our pond, as it were, and I'll do so until someone or something squashes me flat.

There was another time that Bambi had to share the place, a bit. A young woman who lives down the road, Deana, rang me up and told me she'd rescued this buzzard. It had been stunned by a car, the vet thought, but didn't seem badly hurt. He just needed a bit of time to get his wits back. He ended up living in my office, in a cage to begin with. I fed him on roadkill, and on mice caught in various neighbours' mouse traps, and on rabbits that appeared as gifts from people who heard about him and wanted to see him get better. And he slowly began to recover. He

Bambi would always pose for the camera when I took her picture. She would come trotting up and stand still with her ears pricked forward.

Above: *Bambi was very keen on Rich Tea biscuits and liked to share one with me.*

Left: *Given the chance, Bambi loved to lick out the whole biscuit tin. Julie's holding on for dear life here.*

Below left: *Roxy can't decide whether she wants to share her biscuit with Bambi, but it's pretty clear what Bambi thinks.*

Bottom: *My sons, Craig and Stuart. They worked hard to make Bambi's enclosure larger every time we rebuilt it.*

Above: *Bambi was not impressed by the arrival of Tommy the buzzard. She sulked for days until we released him back into the wild.*

Right: *Me with Joe Drewer. This picture was taken when he was very ill, not long before he died. In spite of this, you can see the bags of vegetables he had brought down as treats for Bambi. He was one of her most loyal friends and loved coming to visit her, often twice a day.*

Above: *Smoky the cat, who was a good friend of Bambi. There weren't many animals that Bambi liked to have in her enclosure, but she always welcomed Smoky.*

Below: *I would often lie down beside Bambi's enclosure (sometimes even inside it) and Bambi loved to come and keep me company.*

Anticlockwise: *It took me years to capture Bambi's process of lying down on camera. She had to get the balance of her body just right to be able to go down on her one back leg. It was a work of art every time.*

Above: *Two jackdaws perch on Bambi's back. They were taking the moulting winter fur from her back to line a nest they were building in one of our chimneys.*

Left and below: *When it snowed heavily, Bambi would let it get to an inch or so deep on her back and then shake her whole body so it flew off. Deer don't much like to feel anything on their backs.*

This snowfall was such a surprise to Bambi as it was so sudden and heavy. You can see her wondering what's going on with all the white stuff on the tree.

Below: *When it was icy I used to go out and cover Bambi in straw to be sure that she was warm. Some people said I was mad to do it and deer survive just fine in the wild, but I didn't like to think of her being cold.*

Below: *My cousin's daughter Donna (left) drew this picture of Bambi sleeping (right). It looks just like Bambi did after the vet had put her to sleep so peacefully.*

Bambi came to us on 7 June 1994 and was put to sleep at 4.50 p.m. on 31 July 2006.

made a terrible mess in my office. I was having to cover everything up in sheets, all my equipment, the keyboard and everything. The place was kept shrouded like one of those rooms that have been left alone for hundreds of years, once he started flying around. Tommy, we called him.

And, of course, once he was able to go out and about, he met up with Bambi. She didn't like it one bit if I had Tommy on my wrist. 'Mrrrrr!' she'd go. She didn't want me to have anything to do with him. He was competition. She wanted him out of that office, out of her garden!

Sure enough, the day came when we did take Tommy away – to return him to the wild. All those years ago we'd wished to do just this for Bambi and we hadn't been able to, but we did it for Tommy now, and he was pleased to go. I shall never forget the last look he gave us, from high in a treetop, in the fog, before he flew away.

When we went home there was our beloved Bambi, same as ever. What she should have been jealous of – Tommy's flying away – well, she knew nothing about it. For her it was life as normal. Thank God he's gone, was her feeling about it. She'd behaved more like a cat or a dog. She'd treated that buzzard as an outsider, and the little roe deer as something to leap back from. She was much more content alone in her little garden.

It made her seem like a solitary, stand-offish kind of creature. I was mixing with the wild red deer, up on the moor, probably with her own mother and father among them, and it made me feel guilty, almost as if I was keeping a secret from her. It was sad that she knew nothing of all this. I spent days at a time up there with her kind, observing and photographing, or leading people on safaris so they could see for themselves. Meanwhile she passed those same days in our garden. In a way, I was living her life, or at least something closer to the sort of life she should have had, and she was living my life, a human life. Her days were marked out by what happened between her and me, and her and Julie, and her and Roxy, or any of the other visitors she happened to have. There was a time once, in fact, when they were making a film called *The Passion* in Bishop's Nympton, and somehow it had come about that they were using our bedroom for one particular scene. I happened to tell them, 'I got a deer in the back garden,' and that was that, they downed tools and went off en masse to visit Bambi. Even the catering unit turned up, the chef dressed in his white coat and hat. He frightened Bambi, he did – it wasn't a costume she'd ever seen before. And apart from her visitors, her life was made up with what she'd have to eat: what delicacies I could find in the hedgerows; any vegetables that happened to be growing wrong in Joe Drewer's garden up the road; or a packet of custard creams left on the doorstep by a member of her fan club.

Another hobby for Bambi was organizing her shed. The sawdust was given to her courtesy of Mike Warren, who used to go off in his lunch hour and fill bags of the stuff for us, and by Andrew Aidan, who worked in a carpentry workshop. But she would make her own bed, turning it into a comfortable basin shape. And she kept it very clean, she really was a clean animal. But when the time came to clear out her shed, you could see how much work she'd done because the ground level was three foot lower than it had been to begin with. And she used to tear down the carpets and rugs I nailed up to the walls. I did this because the shed was liable to get cold; its position meant it took the brunt of a cold east wind funnelled in between the houses at the back, and I used to be worried about that. I'd nail up old lengths of carpets and rugs, which she used to pull down again – I don't know if she didn't like the colours, or what it was. So I'd have to go out and nail them back up again. Anyway, it's all in the photographs, just ordinary things: Janet Dibble's twin girls, Marie and Anna, visiting. Lying down with her, sunbathing in my swimming trunks. A family life, in our family.

11 A star on screen

As time went on, we had started going further abroad to show our films, mostly at country fairs and the like, but some way distant from home. So we had a new problem: where to spend the night? We couldn't afford a hotel; that would take all our profits at a stroke.

Instead, Joe Drewer had the bright idea of building a little shed to take to all the country fairs, and so we had this construction made up by Terry Saunders of separate panels, and when we'd arrive at a country fair Joe and I would bolt together these panels, put the roof on, then sleep inside during the night, and during the day we'd open it as a kind of sales booth, for showing and selling the films. I remember one of the early shows Joe and I went to with this contraption was a massive steam rally up near Dorchester. Six or seven hundred acres of ground this show covered. It was a long journey, two or three hours, and when we arrived it was very windy, so we had a hell of a job to stand these panels up without them being blown out of our hands, so we could bolt them together. Eventually we were all set and it was time to go to sleep. But we hadn't reckoned on our slot being hard up against the road, and we hadn't had the sense to bring mattresses, so we were sleeping on the edge of the tarmac, there. It was hell. Joe Drewer called out in the dark, his voice sounding awful, 'Johnny, this is worse than bloody manoeuvres!' He was right, it was. We couldn't sleep. And we were booked in for five nights like this. I found a way of dealing with it. There was a massive beer tent at this show – maybe sixty or seventy beer barrels and cider kegs all lined up. If I could drink enough cider, I knew I'd sleep all right.

We learned our lesson and it wasn't long before we upgraded the accommodation. It was Joe Drewer, again, who spotted that if we could mount the thing on wheels, we wouldn't have to put it up. We could tow it to our spot and be ready to go. It was Joe who organized this new vehicle. It was like a burger van, with a steel frame made by Kevin Boyles, but instead of selling burgers it sold videos and photographs. I used to show the films inside, as well, like a little cinema. 'The Chariot' we used to call it, and me and Julie and Joe Drewer had a lot of fun and games on our trips to the various shows we visited around the country.

Anyway, one day a journalist from the *Daily Telegraph*, a big strong-looking man called Willie Poole, happened to walk past the Chariot, and he took a film away with him, and the next thing I knew, there was a big article in the paper about me. '*Rutting Stags of Exmoor* is a masterpiece,' it said. 'It contains some of the finest film of deer that I have seen; shots that must have involved great skill, patience, discomfort and not a little personal risk.' I was thrilled, course I was. Here was a respected countryman, with his own column in a national newspaper, and he got what I was about. He could see the film was amateur and not very professionally put together, but the point was, I'd got those pictures, and my own excitement and satisfaction was obvious from the narration.

This newspaper article, in its turn, led to a call from one James Cutler, from Yorkshire Television, and he drove all the way down to visit us, to ask me if they could make a film about me and my way of life. This was to be called *The Secret of Happiness*, and six weeks later we were in the thick of it, making that film, being followed everywhere we went by a professional sound and camera crew. It was broadcast as part of the *First Tuesday* series on 2 November 1992, and I watched it in the village hall along with everyone from the village, all our friends and family.

The next day, all hell let loose. The phone was ringing from dawn to dusk, the darn handle was hot in your hands. It was hard to give ourselves time to eat and take a drink of water. It only stopped ringing at ten o'clock that night. The place was chaos, with Stuart and Craig and me and Julie wading through piles of envelopes, and tapes, and so on. It went on all day.

We sold many thousands of copies of tapes. Thousands of letters arrived in the post which all had to be opened and read. Money poured into our bank account. Daihatsu gave me a sponsored vehicle for a year. It was crazy. I was asked to open shows, to talk at functions. It was a giddy time all right.

And yet, within a year or so, it was all over. I'd enjoyed a brief spell of being famous. But, almost as quickly as it had blown up, the storm passed over and we were left back where we used to be.

But I was more committed than ever to my photography, and to making wildlife films. It was the right thing for me to be doing, and the sudden success, however briefly it visited me that time, had encouraged me to keep going. And, just two years later, of course I started to film Bambi, right from when she first arrived.

It happened in bits and pieces, but eventually there was enough material of Bambi to make one DVD that she has practically all to herself. I call it *Johnny Kingdom's Exmoor Orphan*. Rupert Smith, my cameraman on the first TV series, helped me put it together. As I say, it's not like a piece of wildlife film-making, although I do introduce it properly and there's music over the top and so on. But you can't compare it to David Attenborough's *Life On Earth*, that's for sure. The making of this film didn't cost any money. It's our home movies of Bambi, nothing more and nothing less. And that's why it means so much to me. There are bits and pieces of our family life just cobbled together, taken here and there, just when something was happening and the camera happened to be lying around. We all do the same, don't we, capture our family life on film, so we can look at it later and remember how we were? The difference was that every now and again I'd put those bits and pieces into my wildlife films, because people wanted to know how Bambi was, what she was up to, and this was the easiest way of showing people.

Time marched on in the way it does, and all these scraps of home movies began to add up. Almost without knowing it, I was building up a catalogue of sequences of Bambi, showing her life. Eventually, when she was ten years old, I realized there was enough material to make a film of her very own, and Rupert and I went up on the moor and filmed the introductory sequence, and started to cut the material together.

After we'd finished, of course, the film became not so much a way of telling people how we were going forwards with our lives, how we were getting on, in that careless kind of way that we all do, but instead it became a way of looking back over our lives, quite carefully, and with hindsight. Look what happened to him, I'd say. Remember what happened to her, look how so and so has grown up . . . this is what happened. It's a strange feeling to watch our younger selves looking all innocent and unknowing of everything that is to come. It's like playing God with those funny little people on the screen. Yet there's not a thing you can change, not one moment.

Johnny Kingdom's Exmoor Orphan is subtitled

The Story of a Red Deer Named Bambi and it begins with my taking the camera up on the moor, and I show the view down through the Exmoor valley towards Polworthy Farm, where she was found. It's as beautiful a place as you'll find anywhere in the world, especially in the early morning and the late afternoon, when the lowering sun paints every leaf with gold and the shadows move longer and longer across the ground. It's where she would have lived, trotting alongside her mum. I then go and visit the spot where she was caught in the wire, and there is the rack in the hedge made by the deer, you can see it's where they still scramble through, ten years later, and I reach out and touch the broken bit of fence, low enough for a grown deer to step over, and you can see the end of wire sticking out, all rusty now, where Brian cut it, to let her go. It's still there, like evidence in a murder mystery. Next, there is the baby Bambi in the shed, her little corner hurriedly cleared for her, in among all the tools, with a bit of sawdust or straw thrown on the ground and that blue-checked rug for warmth. Her missing leg is still there! That is so odd, to see her with four legs. And around the injury the fur is all grey and flat and damp – ominous-looking. It sends a shiver up my spine to see it.

Mostly, you can see how pretty she was. She stares up at the camera with those beautiful brown eyes, long ears like a hare's, but flicking back and forth, her nose twitching away, as she takes in all the strange scents. So graceful and delicate. Then, here comes my granddaughter Roxy, with a bottle of goats' milk, feeding her. That pink dressing gown looks so odd in the garden. Bambi's leg is missing by now, she has just the one peg leg at the back, but she is already nimble and fast, and she is as big as Roxy is. 'Speak to 'un,' I say to Roxy and she does and you can see Bambi is at home, already. She looks confident and happy, and she's lost that shocked, glassy expression that she had in the shed earlier. Now she looks like she could boss you about, and that's how it was. She grew in confidence very quickly, and certainly she had a very haughty side to her. There's a picture of her eating hazel leaves that I've tied into place along the boundary of her compound and you can see her nibbling the blackberries off the top of a bit of bramble I've cut for her. God, the hours of cutting from hedges! I'd like to add them all up and have them given to me again, now.

And there's Jill, milking the goat in her barn, filling a big plastic bottle of milk for me to take home. During the whole of the period until Bambi was weaned, I'd go down there and pick up first of all one bottle, but then going up to two or three bottles a day, to keep her going. That was Jill Woollacott for you, she was kindness itself. She moved on from keeping goats eventually. Goats only lactate for five and a half months of the year, and so there was a long lay-off time when they weren't earning her any money. Our family's friendship with Jill goes way back. My wife Julie was best friends with Jill's older sister at first, then she met Jill and they became friends. And little did Jill know when this film was taken that she would lose her partner, Roy, who died, and Julie and I would go to the pub with Jill and my best friend Mike Warren, whom I always nickname Brother, until one day Mike asked Jill out, and now they're married, so there you go. I was his best man and I've got a photograph of him and me standing together out the front of his house, and there's a bottle of whisky we're holding on to, and that was before we even left for the church. None of this we knew; we'd never have guessed all that was ahead of us when we made that bit of film.

In the next shot there's me without my shirt on, showing all the tattoos I got from my time in the army in Hong Kong, and I'm kneeling alongside Bambi, and me and the vet are grappling with her, to lay her down on the ground so she can have her stitches taken out. How slender and young she looks. You can see how she lays her head in my lap, how close we've grown in such a short time. I'd say Martin Prior doesn't look much older than this at all, in ten years; he's weathered well.

After that, you can see Bambi scatting about, in and out of her little wooden house, quick as anything, going from a standing start to full pelt in just one leap, and my voice on the soundtrack says, 'Who told you, Bambi, that you'd never cope with three legs, eh?' And it's the strangest sight, to see this one leg, like a gondolier's pole, propel her at such a rate over the ground, round the corners, diving in and out of her hut, quick as anything. Then she's enjoying Julie's roses. Her long tongue goes out and the flowers disappear. I'm not sure Julie can watch that bit. 'Not to worry,' I say to the camera, 'they'll grow again next year.' That hides a multitude of crimes, does that comment of mine. Bambi made sure she was going to punish Julie for being allowed in the house and for being my wife, and she was like an ungrateful daughter to Julie, always just sulking a bit, putting a haughty face on, a little nip here and there, a push, when Julie was so kind to her all the time and never gave up, never stopped all her usual acts of generosity – yes, just like a good mother would treat a daughter who was determined to ruffle her feathers the wrong way, whatever. And then every now and again Bambi would allow herself to love Julie for a while, before the haughty look was switched back on, but don't hold your breath waiting for it to happen again.

Then comes a family to visit Bambi, and this has been a feature of Bambi's life with us – the amount of visitors she's had, especially children. These visitors are feeding her biscuits, which you can see she is enjoying, and there's another shot like that, of Julie holding an empty biscuit tin, and Bambi licking it out, having finished them all. It was her treat, to have biscuits from time to time. We had to be careful because they weren't the best things for her; we maybe should have rationed them a bit more. But then, our pleasures should be taken often enough, after all. What else is life for? As long as it didn't hurt anyone. Biscuits and doughnuts were for Bambi what a pint of cider was to me – something that makes your eyes light up.

Then comes little Roxy, standing in her pink dressing gown, holding up the milk bottle and Bambi sucking away. Strange how such a dressing gown stays in one's mind's eye. I say, 'Look, her spots are going.' That's on 21 August 1994, so you can see how the spots only last a few months on their backs. You can also see, throughout the film, how Roxy grows and how Bambi grows; they were children together. A few shots later both of my granddaughters are there, Roxy and Louise, and they're eleven and eight years old suddenly. Years and years of our family's life have fled past – all sorts of goings on, and I tell you, Bambi's life might not have changed as much as ours but she was our confidante, especially Roxy's, as she didn't have any brothers or sisters. She would have whispered everything into Bambi's ear.

Next, the kids are feeding her a cabbage, maybe one of the crooked ones from Joe Drewer's garden; Bambi is nibbling away and the kids are trying not to get their fingers in the way. And I approach Bambi with a pair of secateurs, saying, 'Her toes have got too long.' I've got less hair; I'm older.

The film jumps forward again, but only a year or two this time – still, it's unnerving when it does that. Like you've been up and down the rollercoaster three or four times all at once. Everyone's bigger, older, and wearing less clothes, the leaves are on the trees and there's green grass underfoot, the hedges swollen with growth, and the sun shines brightly. Here come Roxy and Louise, playing with Bambi, aiming their water pistols into the air so she runs about underneath trying to catch the water. And dear old Joe Drewer from up the road, who was such a big part of her life, takes the straw off her back with a garden rake, of all things, but then wanders up and asks for a brush so he can do a better job of grooming her.

Julie's Ladies' Skittle Team line up along the fence for a visit. There's me, clipping her nails again, and once more it's Christmas, the earth shriven with cold, the plantlife shrunk back, the badgers and the squirrels hibernating. No doubt I came away from trimming her feet with a sore head as usual.

Snow is on the ground, then, and I can remember how it would settle inches thick on her back, and she'd give a big shake and pitch it all off. I know, of course, that deer are hardy, but it made me anxious to see her so unconcerned, lying in the cold. I used to go out there and pile hay on top of her when it was frosty. There's a photograph of her like that, under a pile of hay. I'm sure she didn't need it, but I hope it made life better and she wasn't just humouring me.

In this sequence in the film, when there's snow on the ground, I mention the date, 29 December, in the year 2000. Bambi is six years old, now. She doesn't run around so much. There are more shots of her lying down. She is calmer and more peaceful. It goes without saying an animal can never talk to you about what pain they're suffering, but with hindsight I would say it's likely that she was already feeling the first small twinges of arthritis that would, just like the vet predicted, cause her such discomfort in a year or two. She was slower to get to her feet; there was less racing about. Her demeanour altered, bit by tiny bit.

Next we see a pair of jackdaws standing on her back and she doesn't seem to mind. They were regular guests, after all. She didn't take a blind bit of notice of them. They arrived every spring. She's chewing the cud while they hop from one spot to the next, working away, their beaks pecking at her back. You'd think maybe it's insects they're after, but it's not. As they go up and down her backbone – pecking and pecking – their beaks become more and more crammed with Bambi's fur. They look so comical, glancing around to make sure they're safe. Their beaks are stuffed so full! It's like they're wearing the big type of military moustaches and they're about to give a salute. They'll fly off in a moment and carry the hair to the nests they've already built, and they'll use the hair to line the insides of the nests, binding it together with mud and so on. You couldn't imagine a softer, warmer home than a jackdaw's nest lined with a thick layer of Bambi's fur. As for Bambi herself, she's not worried, even though the pecks looked quite hard. And she won't miss the fur. She's moulting anyway, and she can always grow more.

I like the way that film spins the seasons round so fast, somehow it makes it easier to see how it's the same thing over and over, repeating itself, and while the seasons go round in circles we grow along our straight lines, our individual lives, always onward, but the time passing quicker until it seems, now, as I write this, in my sixties, almost to go as fast and in as big leaps as the film itself. Blink and it's winter; another blink and it's spring.

An Exmoor Orphan ends when Bambi is ten years old, which is a couple of years more than she might have expected if her life had gone as normal, in the wild. As it turned out, she'd have another two years left.

To watch it, of course, I am filled with memories of my family and all the complicated things that have happened between us all, and Bambi has been a much-loved part of that, and I hope we gave her enough pleasure in her life. I think we did. But I can't help feeling sad for her, because she didn't have any contact with her own species, and they are so magnificent, and live in such dramatic and beautiful surroundings. Only a short distance away, they were, and she knew nothing of them.

12 Reaching the end

As we celebrated Bambi's eleventh birthday, I began to notice that I was lying down with her more than before. It had always been a problem for Bambi to lie down. She had to get her weight balanced just right, and lately it was taking her longer than ever. There was no doubt about it: she was senior in years now, and she'd had a life full of fun, but she wasn't the same creature as she had been. For the first time, her morale was low. There was a look in her eye . . . as if she were looking not at you but through you, to the middle distance. She'd get to her feet less willingly, with less spring in her step.

It was the year 2005, and I was making my first TV series for BBC2, so things were flying with me. I was busier than I'd ever been. I'd gone from putting together video tapes in my front room, using two recorders to cut out the boring bits, and taking them round to village halls, all the way to having a full crew with cameraman, sound man, a producer, everything, with the end result broadcast to millions of people on a national TV channel.

But Bambi wasn't doing so well. I called the vet and he confirmed what I already knew. She had arthritis in her hip joint, the same side as the remaining back leg. Martin Prior had always said it would happen eventually. As he explained it to me, the fact she had only one leg at the rear had put everything just slightly out of alignment. Her spine was curved. The top of her thigh bone sat just a bit crooked in the socket. Her pelvis was out of true. Put this together with all the jumping and running she'd done over eleven years, and the joint was worn wrong, and arthritis had set in. She was afflicted with pain. That was hard to bear. I was heavy-hearted about it. We had all rescued her, and given her a very unusual life for a red deer, but a good one, that you could see she enjoyed. However, as soon as that arthritic pain entered the equation, well, this would very quickly tip the balance. My daily thoughts about Bambi began to include this judgement I had to make: how much pain she was in, measured against how good her life was, which in turn was measured against the fact that, after eleven years, I couldn't face up to the idea of losing her.

At first, it helped if I laid my hand on her. This was a gift that came from my old friend Tony Thorne. He was with me one day, long ago, when I had a bad shoulder. He said, 'Come over here a minute, Johnny,' and he put his hand on my shoulder, just rested it there. After a while I felt a burning sensation, like a great heat was going through my shoulder from his hand. And it mended me. True, it did, it mended me. Some time later I was up there again and he said, 'Take my right hand,' and I did, and he held on to my hand, quite tight. 'I know you've got the potential in you, Johnny,' he said, 'and I've always known it. So I'm going to give you the gift, the same gift my mother gave me.' I felt a similar burning and tingling sensation, all through my hand and up my arm, and I believe he did, I believe he passed on that healing power to me, and I used it for a year or so, to help Bambi.

In fact, it was a great sadness to me recently, when that same Tony Thorne died. He was a farmer up at Twitchen and the first time I came across him was many moons ago, up at Terry Rudd's property. I'd built a hide up there, a big one it was, more than 30 foot high, and the bolts and the wires I'd built it with had grown rusty after eight or nine years of use. We were required to take it down because the bit of land that gave us access to it had been sold, and so I went up there with my sons and a gang of mates, and Tony Thorne agreed to come along with his old Fordson Major, because there was no way we could take it to pieces, and we thought we'd just pull it down with the tractor. Very steep ground it was up there, so we had Tony on his tractor facing up the slope, with the rope coming from the tractor and attached to the hide. And he'd let in the clutch and the front wheels would lift three feet in the air, and the hide would bend right over, and then the rope would snap and the Fordson would drop with a thump and the hide would sit right back up again as if it were made of elastic. It was so comical and we were all of us laughing fit to bust, Tony Thorne included. He had to get properly mad with his rope and his tractor before that hide eventually came down. Dangerous, it was, but he did it. Seeing as I was without a hide, then, it occurred to me to go and ask Tony if he had any badgers on his land. I went into his yard – and I can see him now, he was filling up the radiator of the Fordson with water – and I asked him, 'Tony, d'you have badgers on your land?'

He gave me that sideways look of his and said, 'Yes, I got badgers.' I asked him if by any chance I could build a hide near their sett and he answered, 'Course you can, why not?' From that moment we became firm friends, and we both of us loved looking at the badgers and the deer. He had a big stag that lived on his land, a three-and-two-a-top, and there wasn't a rutting season in the last seven years that I haven't gone round to Tony Thorne's three or four times a week to talk about Exmoor's red deer: where they are, what they're doing, which of the big stags is on the up, which heads are going backwards. Two years before he died he was ill in hospital and when I went to see him then I thought he wouldn't come out but he did. Then it was Tony's very best friend, a man by the name of Albert Kingdom – no relation to me – who would say, 'I know I'm going to turn up one morning and find 'un . . .' and sure enough he did. Albert found Tony Thorne dead in his car, his torch in his hand, but he died very peacefully, and there's a lot to say he went there to die, that he knew. Albert went indoors and dialled 999 to fetch the police and the doctor and so on, and the next call he made was to me. Albert lives in the schoolhouse in Twitchen which is close by the cemetery, where his wife

Rosina is also buried. Rosina died after having a premonition that frightened Albert. The subject of death had been discussed one evening and she'd said to Albert in this very particular way, 'I may be next,' and she was right, she died the same night. But I saw my old friend Tony Thorne in his coffin, and it was good for me to have done that, to have seen him so peaceful, because it broke my sodding heart when he died and I haven't the power to go down to Twitchen, yet. It's odd, but when Tony went, the deer went too. They haven't been seen on his land since. And that big stag, the three-and-two: Tony's sister called me to say that Tony would have wanted me to have the antlers of that stag, a magnificent pair they are, and every time they catch my eye they remind me of him, and his great kindness to man and animal alike.

So, just like Tony Thorne taught me, I'd lay my hand on Bambi when she was stiff, and lying down all the time, and I'd just talk to her, you know, tell her she was going to be all right, that she'd be better. And afterwards she'd be up and moving around freely, looking like her normal self. It was like a kind of circle that she and I went round: her off her feet, my laying on of hands, and then her getting up again.

For the best part of a year I kept up with this routine, until I could begin to think: Well, if we can carry on like this, maybe the arthritis isn't beaten exactly, but it's contained. We can live with this.

Towards the end of 2005, however, her episodes of painful suffering began to be too frequent, and the vet Martin Prior came and helped her with cortisone injections, which we did ourselves, straight in her rump. And there were powders she had to be given.

Once more we thought we'd beaten off the pain. There was improvement, hope. I knew it couldn't last for ever, because after all Bambi was in her twelfth year now, and the average age of a red deer in the wild was seven or eight. She was getting on, even if you leave aside the special difficulties she'd suffered.

Once more she began to slide backwards. The cortisone injections began to have less effect.

At this point we heard of a man called Tony Nevin, an osteopath who had a special interest in animals, and who is quite well known. He talked to the vet, Martin Prior, and together they agreed that Tony should come down and treat Bambi. He drove to Exmoor and spent the best part of a day with us. Unfortunately it was wet and drizzling, and I hoped that wouldn't make treatment difficult. I took him out to the garden to meet Bambi.

His view, after he'd observed her for a while, was that Bambi needed to be treated little and often, and so it was important that I learned how to treat her myself, as obviously he couldn't come down twice a week. He explained to me what he thought was going on. 'Normally,' he said, 'a deer divides its weight in the ratio of sixty per cent on its front legs, and forty per cent on its back legs. With one back leg missing, she will have tipped more of her weight forwards, so that will have changed the carriage of her head and neck, which will be held up at a steeper angle than would otherwise be the case.'

I had noticed this already, in fact, that Bambi held her head up higher than the wild deer.

'And also, the remaining back leg will be tucked further underneath her. So she's become a tripod. And that has rotated the pelvis, which causes an imbalance in the muscles of the lower back, causing a gentle but persistent torsion all the way up to the neck.'

He then placed my hands on Bambi, one on the pelvis, the other on the withers, and then he covered my hands with his own, so he could show me what to do, how much pressure, and where. I felt what he was doing: he pulled gently on one hand, and at the same time pushed gently on the other. As he explained, this was to alter the tensions in the muscle chains. He advised this should be done very gently, and little and often, and I should allow Bambi to walk away if she wanted. I'd expected her not to like it, and I'd even told him that I thought she'd move off, she wouldn't put up with it, him being a stranger and so on, but in fact she leaned into us, as if she appreciated what we were doing right from the off. She was the perfect patient; she wanted to do this. For about ten minutes we kept up that gentle pressure on the muscle chains, and after that she wandered off – and she shook the rain off her back.

It sounds like nothing, but I was overcome at the sight of this – because one of the things I'd noticed during this very wet period was that she had never once, not for three whole weeks, shaken the rain off her back in the way that she normally would have done. It was a measure of how stiff she'd become, how much pain she was feeling. But now she did – she shook the water off her back. It was a good sign.

Tony Nevin returned home and I carried on with the treatment that he'd advised. And I had a willing patient: Bambi would come up and almost ask me to do it, time after time. It was making her better.

However, within a few months, she was off her feet even more, and the treatment didn't seem to have any effect. I called Tony Nevin and told him about it, and he said to me, 'Well, that's her telling you, then. She's telling you she's in too much pain to get up. After all, she is twelve years old.'

It was horrible to watch – it hurt me like it hurt her. Yet I couldn't bear the idea of losing her. The vet happened to drop by one morning, and we were both standing in the kitchen here, looking out the back window. 'How's Bambi?' he asked. There wasn't a sign of her.

'She's in the shed, I 'spect,' I answered, and then it was as if she were an actor on a stage, and she'd heard her cue, and it was time to come on. Bambi walked slowly out of her shed and across in front of us. She was stiff, she was in pain, she was low in her spirits.

'That doesn't look too good, John,' he said, and we went outside so he could examine her.

I must just say at this point what a help Martin Prior has been, always doing his very best for us and for Bambi, over many years. On this occasion it was the worst news ever, the news that I couldn't stand listening to. He turned to me and he said, 'John, the time has come, I think.'

'What?' I asked him, although of course I knew what he meant. We could hardly look each other in the eye, it was such a terrible moment.

'I'm sorry, Johnny,' he said, 'but we've got to be

kind. Bambi has to be put to sleep.'

That broke my bloody heart, I tell you.

But he was right. I nodded. I felt a bit dumb-struck, to be honest. It was time for her to go.

'Not right now,' he said, 'but if it's all right with you, I can come back at four-thirty this afternoon, give you time to arrange things.'

With hearts as heavy as lead, Julie and I told the family.

We didn't have much time for making arrange-ments, but it was also a time for remembering. Roxy, my granddaughter, who's sixteen years old now, came out with a whole list of things. 'I remember that time Nanna' – this is what they called Julie – 'made you some Canadian pancakes covered in treacle, and I carried them out to the garden and gave them to you, and you were talking away and Bambi just reached over and stole them right out of your hand, and ran off.'

'Oh yes, I remember that all right.'

'And when I was small, when I was four years old, every morning I got up and I had to be the one to run downstairs first, because I wanted to get her milk bottle out the fridge, so I could be the one to feed her, and you'd follow me out and stand behind me because it was quite difficult for me to lift up the bottle and I was frightened because she was so big.'

'You always wore that pink fluffy dressing gown, didn't you?'

'And the slippers, the bright green bobbly ones with the dogs' faces on the toes.'

'We've still got those, your nanna's got them in a cupboard.'

'And I remember,' said Roxy, 'making Bambi a birthday card, it was green and red and we stuck it on a tree and me and you sang happy birthday to her.'

'D'you remember what you drew, what was the picture?'

'No, I can't. I just remember it was green and red. And there was all those times when me and Louise had water pistols and Bambi would race about, trying to catch the water.'

'That happened a lot, all those games with water.'

'And when Nanna went to work, you'd let Bambi out, and she'd race about and kick mud all over the paths, and trample on all the plants, then Nanna would come back and she'd be mad at you for all the damage.'

'She would holler, for a bit, wouldn't she?'

'Then she'd be all right again.'

'Yes. Always all right again.'

'And I remember talking to her a lot, just sitting in the garden and talking, maybe because I was an only child. "You all right, Bambs?" I'd say, and then rattle on at her about sad things that had happened, or good things.'

'You told her your troubles.'

'Yes. And I remember naming her.'

'Eh? Naming her? She was always called Bambi.'

'Yes, but once, you let me think I could choose her name. And I called her Princess Aurora.'

'Well, I'm blowed. Yes, I remember.'

'After *Sleeping Beauty*, which was my favourite film.'

'Well, I'm blowed, yes, that's right. Only for a few days, but how about that, she was called Princess Aurora.'

The arrangements preoccupied us. We had to decide where we were going to bury her, and how we were going to transport the body, how to carry her out of the house. We wanted to do this properly, do it well.

It wasn't a problem, deciding where her last resting place would be. We knew exactly where. It was almost as if it had happened on purpose.

A couple of years before, I'd been standing at my stall in Tiverton market, selling my videos, when I fell into conversation with Edna and Mervin Cowling, a couple who were interested in wildlife. They owned a stretch of woodland near our village here, Bishop's Nympton. The conversation turned to the subject of badgers. 'You like badgers?' they asked and I replied, 'Yes, I do.' They told me of the badger sett in their woods and they invited me to go and film there, if I wanted to.

I went there, and sat and waited for the badgers, and, since it was May, I found myself in the middle of a sea of bluebells, with the late evening sun slanting down through an oak tree above. It was like being in heaven. You couldn't imagine anything prettier.

Then – to cap it all – two badgers came out, and then the sow, the female badger, was followed out by three small cubs, only eight to ten inches long. And the cubs began to play, rolling around together and standing up on their hind legs, pushing and wrestling and tumbling on the ground. These magical creatures were unaware of our presence and, like always, it was a God-given privilege to observe them.

The next time, I went with my second cameraman for the BBC2 series, Rupert Smith. We filmed five badgers altogether, playing. One nearly came close enough to touch. Rupert was behind, filming.

This lovely piece of woodland was owned by Edna and Mervin Cowling; they had it for their own pleasure, without any intention of developing it or building on it, apart from the track they made so they could get around it on the quad bike.

I asked if I could build a hide; and they said yes. It fitted in with what they wanted for their land – a place for wildlife, and for people to see wildlife. It also suited the purposes of the BBC2 series; they could film us building the hide and with their contribution from the budget we could make a better job of it.

This new hide was wonderful. I could sit up there in my little house in the trees and watch the badger cubs, or the birds, or the deer, to my heart's content.

And then, on top of all that good luck, things went one step better. I received a telephone call, and it was Mervin Cowling, the owner of the woodland where I'd built my new hide. He'd heard of a disappointment I'd suffered recently, in not getting a plot of land that I thought I was going to buy – the sale had fallen through. And just to put this in perspective, for all the miles I've trekked across Exmoor, for all that I know every inch of the river Bray, for all the sixty-seven years I've been alive, I've never owned a single blade of grass of it, nor my father before me, nor his father before him. For the most part, our family have been chased off other people's land by bailiffs armed with sticks. All the hides I'd built up until this point – and there's been a fair few – have been allowed thanks to the generosity of various farmers who've let me use their land because they approved of what I was doing and were interested in animals themselves.

You might say that I'm the very opposite of landed gentry.

But, when I was lucky enough to have the success of the television series and I had some money in my pocket, at last, I could afford some acres of my own. I thought I was buying some land, the offer was accepted, but suddenly the sale had slipped between pillar and post. I felt sore about it.

Anyway, Mervin called and he said to me, 'How about, Johnny, you buy our piece of woodland off us?' They had, apparently, always thought they'd sell it to me and Julie eventually, but when they heard about our disappointment they decided to bring their decision forward.

This was a bigger piece of luck; this was amazing. Not only was it the most beautiful stretch of woodland, but it's where I'd already built my best hide ever. I leaped at the chance, and there it was – suddenly, I had a bluebell wood of my very own. The hide I've built can stay up for ever. There's no one who can chase me off, no one can come along and say, 'Johnny, I'm sorry, you have to take it down, move on now, please.'

It was a wonderful feeling; it was kind of like coming home. The land is bigger and better than I could ever have imagined owning. I can walk those acres; and every tree and bramble and blade of grass growing – and all its creatures breeding and feeding and fighting – are on my patch. It's such a pleasure to me. Recently I moved the Chariot down there – that same shed built by Terry Saunders that we slept in and showed our films in, that's been sitting on a patch of driveway over the road, retired for years. I've turned it into another hide, believe it or not, and I put it alongside a big pond I've dug out; around three parts of an acre in size, and 7 feet deep, fed by water from the natural spring from the hill, clear as anything. From inside the old Chariot, that I had so many adventures in, now I can watch the ducks fly in through the trees and settle on the water. I've grown buddleia plants all around it, so in just a year it's become almost invisible, and of course the butterflies love it and they've started to come. And I can say to the Chariot as well, after all its hard work, rest in peace. It won't be going anywhere.

All this meant there was no question where Bambi should be buried. On our own land. Right there, just inside the gateway, is a patch of ground where the sunlight always falls, and there's a hazelnut tree, one of Bambi's favourites, growing above it, providing dappled shade.

It was around twelve noon on 31 July 2006 when my son Craig and I went down there to dig the grave. We had a mini-digger with us, lent by Paul Suntah and Nick Thorne. Craig worked the digger, its long hydraulic arm scooping out bucketful after bucketful of earth, while I stood some way off and watched. It wasn't the easiest ground, with rock quite close to the surface. Craig dug down around 5 feet.

Neither my sons nor I are strangers to gravedigging. We were brought up to dig graves the old-fashioned way, using a Devon shovel, a twelve-prong stone fork with the handle shortened a bit to fit in the confined space, a pickaxe, a saw to cut through any tree roots, a sledgehammer to break up stones. Gravedigging has been the Kingdom family business since I don't know when. And even before I was old enough to help with the digging, on winter evenings I'd hold the Tilley lamp and move from spot to spot around the edge of the grave so they could see what they were doing.

Many, many years after that, I would be in that very same churchyard, spending two or three hours – quite senior in years myself – breaking up the earth with pick and spade, lifting it out with the shovel, digging with my own hands my father's grave.

And the grave where we buried him is the same grave where my nephew little Paul is buried. He'd died many years before, aged three years old. They chose, as his last resting place, that bottom corner of the graveyard which Father and I stepped across on our way to work at the quarry. For the next forty years, every day, Father'd go out the house, turn right in the lane, up to the churchyard, through the wrought-iron gate, give a nod to Paul's grave and offer a prayer for the soul of that poor little lad, as well as for the soul of poor little Matthew, another grandson, who lies in the next-door grave, before stepping over the bank and heading down the hill. And then he was buried in that same spot himself.

And, years later, sharing the same grave, came my mother, his wife. Both my parents died during the period that Bambi was living with us. Mother had died in her sleep peacefully, and we'd promised that she'd rest alongside Father, when the time came. We had the same casket made as his, with the same brass plate, to match. I returned to that same corner of the graveyard with my spade, pick, and shovel. She had reached the age of eighty-six, near enough the same as Father, so you could say they'd had their fair share of years.

I recently worked out roughly how many graves I've dug in my life and it must be near enough three thousand. Among the children in my class at school, sitting next to me, doing the same lessons, playing football like me, I've buried seven of them.

To be a gravedigger means you think about life in a different kind of way. To watch people go through the process of death, its rituals, its grief and sorrow, its permanence, has given me a greater reverence than I otherwise would have had, I think. Has it been because of the gravedigging, I wonder, that I've had so much more than my fair share of death, around me? There are some people who go their whole lives without seeing a dead body – that is, before the deaths of their parents – yet I seem to have that black figure, with his scythe, always walking alongside, always putting me in the way of death, close to it. I found the body of Herbert Thorne, my boss at Sindercombe, who'd died of a heart attack – and all of one cold, cold night his faithful dog stayed by his master's body, before I came along the next morning and discovered him. Another time, I walked into the Tiverton Inn only to be asked to see if Marion's husband, Dick, really was dead, sitting upright in a chair, his chin lowered on his chest. I could see he was dead, before I even checked for a pulse.

I've had so many dealings with death, over the course of my lifetime, that I can look it square in the eye and keep it off for a good many years yet, I hope, with a good joke and a drink of cider and with the help of our friends and some more work to do and our grandchildren to watch growing up.

Gravedigging has been a tradition of the Kingdom family for four generations: my grandfather, my father, myself, and my sons. Our retirement from the family business came about as a result of the mini-digger. There's no need of so much manpower any more. So we'd told ourselves that we'd dug our last grave together, Craig and I, but perhaps we can say that death always has the last word on any matter, because here we were, my son and I, digging another one.

For Bambi. In our woods, our land, that was our home. It seemed like I had such a weight on my shoulders, the weight of all three thousand of those graves I'd dug during my life, while we dug Bambi's grave.

It didn't take long with the mechanical digger. We didn't have to swap round, didn't have to break a sweat. The job was done in under half an hour. A grave was dug not for a human friend this time, but for an animal one.

We parked the digger out the way, off to one side, and drove the few miles back to Bishop's Nympton. The house was very different from usual. It was quiet and sombre. Waiting there was my other son Stuart and his daughter Louise, and various other friends and family. Martin Prior, the vet, turned up as he said he would, at four-thirty. We went out through the house to the back garden. Bambi was lying down, leaning against the wire. I went into her pen and she got to her feet, which wasn't what I wanted. Ever so gently I pulled her down again. Maybe it was just me, but I had the feeling she knew this was the end. I had her lying on the ground and I tucked myself right up next to her, so we were lying together, just like when I first visited her after the operation, twelve years before, and she stopped crying. Once again she was soothed as I stroked her. It was important for me not to be upset or stressed, because that would transmit to her. If I was calm, then so would she be. The first injection went in her rump, and it made her sleepy. Already her head felt heavier in my arms. I talked to her the whole time: 'I'm being kind, you must understand that, Bambi, you must understand what we're trying to do, you're in pain, too much pain, and it will only get worse, and so we're being kind to you . . .' and so on. She didn't understand the words, but what she did understand, I've no doubt, is the tone of my voice. After all, we knew each other very well.

Now that Bambi was sedated, Martin Prior gave her an injection of Pentobarbitone in her neck. This is a powerful barbiturate and it relaxes every part of the body until it stops working – the heart, the mind, everything just slows down inch by inch.

She was breathing heavily, and her head was against my head. I could hear her steady, deep breathing as she slept. I found myself copying her. For every inward breath, I took one too; when she breathed out, I breathed out. Somehow, I wanted to make this journey with her, just as far as I could, right to the very border to the unknown. She would only leave me when she made that last step across to the other side, whereas I would stay on this side for a while longer.

I did. I went with her all the way. I held her in my arms until she died.

It broke my heart. To write about it now takes me

right back there. I'm in my office, and just outside, a few feet away, is the garden where she lived for twelve years. It's an ordinary garden, same as anyone else's. After her death the boys and I worked like slaves, quick as we could, to put the garden right, to take down the fence, to make the lawn, because we wanted to move on, escape from that feeling of emptiness.

But to go back to the moments just after Bambi died, the vet and both our sons thought it would be best if Julie and I went off for an hour or two. It's never an easy thing, to have to watch the handling of the body of a loved one. It is shocking; it brings it home to you that they are gone, they will never look you in the eye again. The dead weight of a body is quite different from anything else and it's a painful sight. They would attend to the moving of Bambi out the front and into the truck and we should come back in an hour or so.

We went back into the house, all of us really low, afflicted by terrible sadness. And do you know, I happened to glance at the wall in our kitchen, and there was a pencil sketch of Bambi hanging there, drawn by Donna, the daughter of my cousin Barry Bailey and his wife Jenny. I almost felt my knees go, because that picture could have been of Bambi as she lay right now, how we'd just left her. Donna is a schoolteacher, and she loved to go out to the garden and sketch Bambi. And she'd drawn Bambi asleep in this picture, and she'd got it dead right, Bambi was asleep for evermore.

Julie and I got in the car and we left Bishop's Nympton and drove to the top of Four Crossways. From here we got out and walked towards Mollan Moor. This was the home of the red deer, Bambi's spiritual home, and ours, too. We could see for miles. We didn't talk much. Both of us knew what the other was thinking. We were remembering Bambi, and we were adding up all that had happened in our lives since Bambi had been in our family. I'd seen both Father and Mother into their graves, and Julie's mother also had followed her father over to the other side. We'd seen our granddaughters grow up: they'd turned from toddlers into teenagers in front of our very eyes and Bambi had given such a lot to their childhood. I think that keeping an animal is important for children; it teaches them kindness and responsibility and brings a sense of God's earth, of God's magic, into their daily routine. We'd seen our lives transformed in terms of income and work: with the first television series about to air on BBC2 we were further ahead in life than we ever dreamed was possible. And quietly, unknowingly,

Bambi had lived alongside us through all of this. And she'd given us so much.

'You'll get your garden back,' I said to Julie.

'Yes, I s'pose I will,' she said, and she sounded so sad. This is the woman that I had my eye on when she was just fourteen years old, and here we were fifty-odd years later, still walking alongside one another, with the heights of Exmoor all around us, so lovely, so unchanging.

Every now and again we checked our watches. When an hour had passed, we walked back to the car and drove home again.

They'd carried Bambi's body to the trailer which was hitched to the back of my truck. And there it was, covered with that same blue-checked blanket that she'd always had, for all her long life, to keep her warm, to see her out the other side of any illness, but not now. It did put a crack in your heart, to see it put to this final purpose, because there was no warmth in her any more.

We drove slowly to our land, to the gateway, and turned in. There was no rough handling of her; she was lowered into her grave so gently, while Julie and I watched. 'Make sure her head is facing the gate,' I called. I wanted her to see us coming and going, as it were. Every time from now on that I'd come through into our land, I would give a silent greeting to Bambi, just to acknowledge her, and I wanted to have the sense that she was facing me.

When she was in there, her final resting place, the digger was brought alongside and the grave was backfilled.

Another grave. I've had too many. Is it that, which makes me quick to find a laugh, to have a good time, and quick to take down a pint of cider, and latch on to the doings of living creatures, the magic of nature? Maybe it is.

We dressed the top of the grave with clats of earth, and it was an easy job to go into the woods and find a hundred bluebell bulbs and move them to this patch, so in spring Bambi's grave is a carpet of colour, always sunny, but yet cool and quiet, as beautiful as you could want, as lovely as the first day I came to these woods to watch the badgers playing.

There was a cross made for Bambi's grave, given to us by Steve and Jane Westacott, which was engraved and stands for her memory. We took our leave of her, giving her our last, most private thoughts, and stepped back from the graveside, and walked away.

Even now, Julie tells me, she finds herself walking around the very outside edge of our lawn, where you used to have to go when Bambi's pen was there. Julie has to remind herself she can alter direction, walk straight across. We used the posts and the wire up in our woods, to make the keeps for some trees we'd planted. And Bambi's shed was cleared out, of course. All the sawdust bedding taken out, the floor put back to earth, as it was before. The walls hosed down. The lawn was made good, the bare patches sowed. We dug the pond in the bottom corner, and now it's home to a lot of frogs and toads, let me tell you. We put stepping stones across the grass. There are more bird feeders than ever. We planted shrubs. It has become a garden again.

As for that shed, which was Bambi's home for twelve years, well, it's gone back to being a garden shed again, only a lot tidier. It doesn't have Craig's spacehopper or Roxy's toy wheelbarrow. We've gone beyond that stage. It's a grown-up home for the lawn-mower, for the rakes, for the spades and forks and so on. And there is a brass plaque screwed to the wall, given to us by Tony Panting, engraved with Bambi's date of birth, 7 June 1994, which is when she was found, only days old, and the date of her passing away, in my arms, 31 July 2006, as well as the words, 'Remembering our lovely friend Bambi'.